# INSTRUMENT FLYING GUIDE

By

ROBERT T. SMITH

Author of CLASSIC BIPLANES and AIR TRAFFIC CONTROL

MODERN AIRCRAFT SERIES

TAB BOOKS

Blue Ridge Summit, Pa. 17214

THIRD PRINTING—APRIL 1978



Printed in the United States
of America



Hardbound Edition: International Standard Book No. 0-8306-9933-3

Paperbound Edition: International Standard Book No. 0-8306-2217-9

Library of Congress Catalog Number 64-25008

# Contents

## ACKNOWLEDGEMENT

Bendix, Collins, and Narco were very generous in supplying photographs of their radio equipment, and my thanks to them. Thanks also to Bill Robinson of Beech Aircraft and Jerry Kell of Cessna for your good help and excellent photographs. And to Mrs. Cecil Gilham, my special thanks for allowing me to use Fulton Air Service facilities and airplanes, and for being so very cheerful and friendly.

Robert T. Smith
Smyrna, Georgia

# 1. What Is Instrument Flying?

When we first learn to fly, we are taught to fly the aircraft by controlling its *attitude* with respect to the earth's horizon. When the horizon is not visible (for any reason), it is impossible to control the attitude of an airplane unless the airplane is equipped with suitable instruments, and the pilot knows how to fly the aircraft using those instruments. Thus it follows that a pilot who is not trained to fly instruments may easily meet with disaster if he allows the aircraft to enter conditions where the earth's horizon is not visible to him.

**Not As We Feel**

Our sense of balance is of no use to us under instrument flight conditions (commonly referred to as IFR or Instrument Flight Rules conditions). We maintain equilibrium or balance because we can touch things, feel pressure and tension, have the balancing organ of the inner ear, and because we can see. If any one of these is lost, the ability to maintain one's balance is seriously impaired. Sight is by far the most important.

The balancing organ of our inner ear is a motion-sensing device consisting of three small canals, each arranged in a different plane. The canals are filled with fluid into which project tiny hairs. As we move our head, the fluid moves thus causing the tiny hairs to move, sending a signal to our brain which tells us we are experiencing motion in one plane or another. Accelerating to a faster speed, slowing down, turning, tilting, all cause the fluid in the canals to move resulting in signals being sent to the brain. These signals tell the brain the direction and intensity of the motion. Sometimes we experience motion in several directions at once, for example when we roll an airplane into a diving turn. We get rotation motion from the turn, and up and down motion from the dive. If, in addition, we gain or lose speed as we enter a diving or climbing turn, we add the motion of acceleration or deceleration. From all the signals detected in the canals of the inner ear, the brain determines the final resultant motion, and the approximate intensity of it.

An airplane is free to move about in space, accelerate and decelerate. Because of this, the final resultant signal from the inner ear canals is sometimes erroneous. For example, if we decelerate while turning, the inner ear senses a reverse turn as the final motion. However, if we can see the earth's horizon, we can see that we haven't entered a reverse turn, and so we disregard the signal from the inner ear canals. The process is automatic *when we can see the earth's horizon.*

Under instrument conditions, we cannot see the earth's horizon, so a deceleration while turning would cause us to sense a reverse turn. We would try to correct it by applying control in the direction we are already turning. This would only steepen the bank of our turn. Since our inner ear has given the brain a false signal, and we cannot see the ground, we have nothing except the sense of touch and feel to help us check the steepening turn. The sense of touch (postural sense) is not as refined as the sense of sight or the inner ear, and responds to the total result of all forces acting on the airplane.

If these forces (centrifugal force and gravity) are in good balance, our postural sense would tell us we were still in the original turn which would only serve to confuse us more considering the brain's reverse turn signal from the inner ear! So, the sense of touch and feel is of little use to us under instrument conditions. All we have left is the sense of sight, and we're under instrument conditions in a steepening turn. As the bank increases, the nose drops, and the airspeed increases. This airspeed increase is not rapid, so the inner ear may not catch it—it usually won't, and even if it does, it may not interpret it as a speed gain since we are also losing altitude which gives us a downward motion.

If our aircraft is equipped with at least an airspeed indicator we can use the sense of sight to observe the airspeed indicator and note that it is increasing. To decrease airspeed we should raise the nose. However, we are now in a steep turn, so when we attempt to raise the nose all we accomplish is a further steepening of the turn. Back pressure on the stick causes us to sink down slightly in the seat and our postural (touch and feel) sense tells our brain we are climbing because, as we enter a climb, we sink down in our seat. Now, our brain tells us to lower the nose, but our eyes see the airspeed increasing—what are we to do? Frankly, at this point it doesn't make much difference *what* we do—we are in a steep, diving turn (a spiral), and none of our senses can get us out of it, particularly if our sense of sight has no method of allowing us to see that our wings are not level. We are in what is commonly known as a "graveyard spiral" so named because the end result is usually fatal.

**In the Beginning**

The first airplanes had no instruments of any kind, neither flight nor engine instruments. As the technical art of aircraft construction advanced, instruments were added to monitor the engine's condition, and to tell the pilot how fast he was going, how high he was, and in what direction he was flying. The altimeter, airspeed indicator, and compass were the principal flight instruments in use during World War I from 1914 to 1918. However, none of these instruments told the pilot when his wings were level.

**Pilot of this Curtiss pusher biplane sat out in the breeze with no instruments of any kind. Few such pre-World War I aircraft had any instruments.**

Barnstorming pilots carrying passengers in Jennys and Canucks had no reason to fly under instrument conditions, but when they started carrying the mail, it became necessary to penetrate instrument weather on occasion. Zipping through a small puff of cloud posed no problem, but flight in clouds for more than a few minutes generally led to disaster. Unknown to the pilot, the airplane would enter a gentle turn which would cause the nose to drop, increasing the airspeed. By the time the pilot noted the increasing airspeed, he was in enough of a bank so that just pulling the stick back served more to steepen the bank than it did to raise the nose. The result was that the airspeed continued to increase even though considerable back pressure was applied to the stick.

**Amelia Earhart's first airplane, this 1928 Avro 'Avian,' had basic instruments for VFR flight: compass, air-speed and altimeter.**

Few pilots recovered from this resulting "graveyard spiral" unless there was enough room under the clouds to return to straight and level flight before the wings were torn off. If he had had that much room, the pilot would have gone under the clouds to begin with! The graveyard spiral persists today, kept alive by VFR-capable pilots who "push" marginal VFR conditions by trying to get through when the cloud ceiling (height of clouds above the ground) and visibility are so low the earth's horizon begins to disappear. The result of attempts to fly VFR when the weather is marginal for VFR are usually fatal.

## A Solution

Designers and pilots of the early '20s tried many schemes and devices in an attempt to fly in instrument conditions and remain right side up. The various devices and schemes had one thing in common—none of them worked. One early airmail pilot had a most delightful idea that *almost* worked! Like other pilots of the period he had his own pet idea on how to fly under instrument conditions. (They didn't call it "instrument conditions" back then for the term was unknown to them.)

This airmail pilot surmised that if he had a transparent container filled with a liquid to observe when flying in the clouds, he could tell by the slant of the top surface of the liquid whether or not the airplane was level. So, on his next night mail run, he brought along such a container filled with liquid. Sure enough, the weather was not the best, and he was soon flying in and out of small, broken clouds. He pulled the container out of his pocket, and held it up for examination. The container he had selected was a bottle and he quickly noted that the liquid was so far up into the neck of the bottle that its top surface was difficult to observe. He deduced that it would be necessary to remove some of the contents of the bottle in order to have a larger top surface which would afford a better reference.

He first thought to open the bottle, and pour a slight amount of the liquid overboard, but was afraid the slipstream would suck all of the liquid from the bottle. He next considered dumping some of the liquid into the bottom of the airplane, but decided that might be detrimental to the plane's fabric covering. Finally it came to him that he would be forced to drink the necessary amount of liquid from the bottle, and having had this revelation, he proceeded to lower the top surface of the liquid in the bottle to what he considered an acceptable and usable level.

Each time he entered one of the puffy, moonlit clouds he would hold the bottle up to examine the level of the top surface of the liquid, and each time he examined it, he decided that the liquid was still too high, and would be more helpful to him if it were lowered. Since he had already decided that trying to pour the liquid overboard might cause him to lose it all, that dumping it into the bottom of the cockpit might damage the airplane's fabric, he was forced to continue drinking the liquid in the bottle in order to get it down to a level where it would be usable as an instrument for a pilot caught in clouds.

The liquid-lowering procedure was necessary more times than he had at first thought it would be. For one thing, the bottle was not made of perfect glass, and he never seemed to be able to get the liquid level to a point where there were not some distortions of its top surface by discontinuities in the glass. Also, it became apparent that the lower the liquid level in the bottle, the more stable its top surface became which would be highly desirable while attempting to fly through clouds by observing the liquid level. Thus it was that he drank the contents down until only a bare half-inch or so remained. At this point he broke out of the scattered clouds into the clear, and decided he had no further need of the liquid level, so drank the remaining half-inch.

He then threw the empty bottle overboard, and it struck one of the stabilizer brace wires causing it to break. This allowed a severe vibration to develop in the stabilizer, so he elected to land immediately. A brightly moonlit cow pasture was picked out, and a more-or-less successful landing made. It was *less* successful than usual because he hit the ground hard, washed out the landing gear and tore up a wing tip. But, it was *more* successful than normal because he was unable to stand up and climb over the side of the cockpit, and the airplane did not catch fire which fact saved his life. Since the airplane did not catch fire, and he was many miles from habitation, he decided to remain in the cockpit, and soon fell sound asleep. It may be presumed that he burped occasionally as he slept.

### A Better Solution

The three instruments available at that time—the airspeed, altimeter, and compass—did not give the pilot sufficient information to enable him to determine when he was entering a graveyard spiral. The compass might have, but it was only a basic magnetic compass (similar to what we still use today) and it oscillates and swings so badly when an aircraft is turning that it is of no use in detecting the direction and intensity of a turn.

Since the graveyard spiral involves a turn, it seemed reasonable that if a pilot could avoid a turn he could likewise avoid the graveyard spiral. In order to turn, an airplane must bank. The first attempts at preventing the graveyard spiral involved the use of a bubble level to show the amount of bank. The only thing wrong was that the bubble remained on center in a turn! The reason, of course, being that in a well coordinated turn centrifugal force balances gravity which causes the bubble to remain in the center and show no bank whatsoever! This device was later developed into the slip indicator, and now is used to show the quality of a turn. It is usually combined with the turn needle in the turn and bank indicator, and is known as the "ball bank indicator."

Pendulums and other gimmicks were tried, with no success. They too suffered from the effects of gravity and centrifugal force. It was not until the principle of a gyroscope was applied to the problem that it was satisfactorily solved. A gyroscope is a spinning wheel, usually a wheel that is heavy for its size, i.e. a high density, rotating mass. A gyroscope has two principles that have been applied to aircraft instruments: rigidity in space, and precession. The first gyroscopic instrument used precession. When a force is applied to the rim of a spinning gyroscope (commonly shortened to "gyro"), the force will move forward in the direction of spin, and in an amount of 90°, and be felt at that point against the gyro, causing it to tilt.

It was learned that a gyro mounted on a simple gimbal ring could be used to indicate when an aircraft was turning by having a needle geared to the ring so that the needle would deflect any time the gyro was caused to precess. The force that caused the gyro to precess was any turning of the aircraft. The gyro, if spinning fast enough, would cause the turn needle to deflect any time the aircraft was turned even a small amount.

On most early turn instruments, a ball-bank indicator was installed to show quality of the turn being made. The ball-bank indicator is nothing more than a carpenter's bubble level. In a properly coordinated turn, the bubble would remain exactly centered, just as it is in straight and level flight.

With practice, pilots learned to fly safely in clouds for prolonged periods, using the turn needle to keep the wings level by preventing a turn. Thus, the sense of sight was furnished a somewhat rudimentary "artificial" horizon to observe in order to prevent the graveyard spiral.

**Where Am I?**

Although the turn needle allowed a pilot to fly in clouds, it could not tell him where he was once he entered clouds. In order to accomplish this feat, something was needed that would give the pilot his position relative to a known spot on the earth's surface. At first, lines of rotating light beacons were set up at ten-mile intervals along the airmail routes to guide airmail pilots at night.

At night, in good weather, the rotating light beacon was an excellent navigation aid, and from 1926 to 1929 airway mileage with beacons installed rose from two thousand miles to ten thousand miles. Though they were useful only at night, the beacons were at least a first positive step toward a ground-located navigation aid for pilots. Even in bad weather they were sometimes useful. Many the airmail pilot who struggled from airway beacon to airway beacon on dark and stormy nights. Some made it, some didn't.

In 1925 experimental work was started on a radio beacon. By 1927 the radio beacon had taken definite form as a radio station that would transmit signals in such a manner that four "beams" were projected outward from the station. In order to fly these "beams" pilots listened to an audible signal transmitted by this new radio beacon. Within a few years the radio beacon was being installed along the airmail routes to enable pilots flying instruments to find their way. Thus was born the airway system in this country. Pilots flew from one radio beacon (or "range") to another along the airways.

**First Step To Greatness**

By the late '20s the airmail lines had expanded their services to include passenger carrying, and the scheduled-flight airlines were born. From radio range to radio range they hauled their passengers and carried the mail. By the early '30s a few executive airplanes were beginning to fly instruments. One of these was a Beechcraft 17R flown by E. F. Ross for the Loffland Brothers Company out of Tulsa, Oklahoma.

In those days there was no air traffic control. The only people flying in bad weather were the airliners, and some few executive airplanes. When Eddie Ross started flying instruments in the Beechcraft 17R, he made a deal with the airlines to listen on their radio voice frequency so he'd know their altitude. With this coordination, he was able to fly at altitudes that the airliners weren't using with their tri-motor Fokkers and Fords, and twin-engine Curtiss Condors. Thus was maintained what is called "traffic separation", i.e. two planes did not fly at the same altitude at the same spot at the same time! It is in this manner that airplanes are prevented from running into one another while engaged in instrument flight.

**Instrument panel typical of the World War Two period. This is a Beechcraft model G17S panel built in 1946 using World War Two instruments. Notice the turn and bank instrument is at top of the panel with artificial horizon given lesser position of importance in lower lefthand corner. (Beechcraft Photo)**

In the very early days of instrument flying, the only navigation aid was the four-course low-frequency radio range. It transmitted four courses which the pilot recognized by a constant tone in his radio headset. If he drifted off this course, he heard the Morse code for either an A or an N, depending from which side of the course he had drifted. Development of this radio range had begun, as we've already noted, in 1925. By the early '30s enough of them had been set up to make regular instrument flight feasible.

Most major cities had a low-frequency radio range located near the airport so that a pilot could make an instrument approach using the range. Unfortunately, due to the wide (three degrees under ideal conditions, more under poor conditions) range courses, a really precise approach was not possible, and it was necessary to have a good ceiling and decent visibility under the clouds to make a range approach safely. Something on the order of a ceiling of 500 feet, and a visibility of one or two miles was considered minimum.

As radio aids to navigation were improved through the years, the major problem continued to be the approach phase of a flight—that phase from the navigation aid to a point from which a landing could be made. With the low-frequency range the ceiling and visibility needed to be about 500 feet and one or two miles, *if* the range was located near the airport. If it were several miles distant, the ceiling and visibility requirements increased. The later introduction of ILS (Instrument Landing System) and GCA (radar Ground Controlled Approach) permitted ceiling and visibility of around 200 feet and one-half mile. No system even today is capable of landing an airplane with zero ceiling and zero visibility. This remains today the major, unsolved, problem of instrument flying.

**Cessna Skyhawk panel for 1973 shows orderly layout with flight and navigation instruments on left panel, radios in center, and engine instruments on right side. Instrument panel layout plays an important part in pilot's crosscheck or "scan." (Cessna Photo)**

## 2. The Instruments

It is not necessary to be an engineer to fly an aircraft with reference to instruments. Neither do you need to have an engineer's knowledge of the internal mechanisms of flight instruments. But at least a basic knowledge of how flight instruments are constructed is very desirable—it will increase your confidence in them, and tell you what their limits are. We will discuss them in the following order:

1. Magnetic Compass
2. Airspeed Indicator
3. Altimeter
4. Vertical Speed Indicator
5. Turn and Slip Indicator
6. Gyro-Stabilized Heading Indicator
7. Artificial Horizon

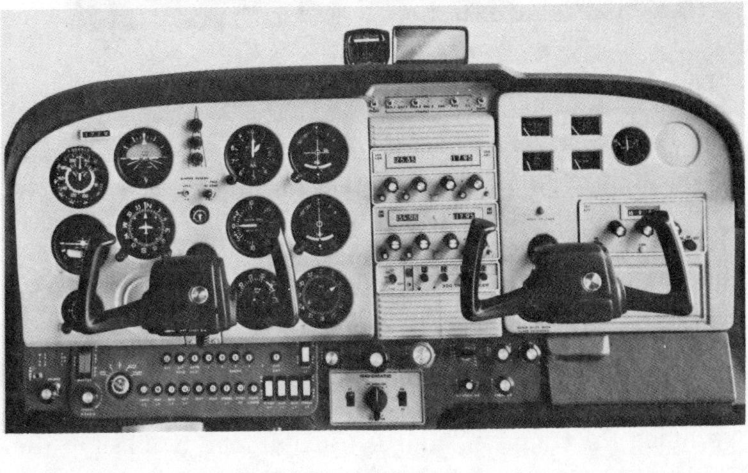

**1. Magnetic Compass**

Many centuries ago ocean navigators learned that a needle made of iron would always point in the same direction if mounted so that it could rotate freely. From this simple needle the magnetic compass was developed for shipboard use, and when airplanes came along, the ship's compass was adapted to aircraft use. The reason the early iron needle pointed in the same direction all the time was because the earth is a giant magnet, and the needle was simply aligning itself with this giant magnet's lines of magnetic force, or flux.

**Typical layout for Cessna 310 panel in 1973 light twin. (Cessna Photo)**

The basic aircraft magnetic compass consists of two small bar magnets mounted on a jeweled pivot, and suspended in a fluid to dampen their movement. Around them is placed a card marked in degrees. Directly above them are two very tiny magnets mounted so that their position may be changed by adjusting screws. By moving the adjusting screws it is possible to correct the magnetic compass for "local" magnetic disturbances such as aircraft radios, or the metal of the airplane itself. The process of making this adjustment is called "swinging the compass," and is usually done by placing the aircraft on a compass rose marked on the airport ramp, and comparing the magnetic compass reading to the direction indicated by the compass rose. If there is a difference, the adjusting screws are used to align the magnetic compass with the compass rose.

A more accurate magnetic heading than that provided by the simple magnetic compass is obtained with a flux-gate compass. The flux gate is mounted in a wing tip, or some other remote place so that local magnetic disturbance is at a minimum. Magnetic direction is determined by the electrical energy of the earth's flux lines as measured by the flux valve and amplified to usable voltage by a compass amplifier. Direction is then displayed on a suitable dial in the cockpit. Because of its complexity, the flux-gate compass is expensive, and is normally found only on large or costly aircraft.

**1973 Beechcraft Sierra panel is neat, uncluttered. (Beechcraft Photo)**

The flux-gate compass is not susceptible to as much oscillation as the plain magnetic compass, and since it is remotely located, it is generally more accurate. If a directional gyro is installed in the flux-gate compass system to stabilize the directional signals during turns when the flux gate is not entirely accurate, then we have a gyro-stabilized compass which combines magnetic heading, and the stabilized turning indications of a directional gyro. Such a system is the ultimate in compasses.

**Beech 125 corporate jet panel has flight director system, radar, and advanced radio grouping. Copilot's side has a full flight group also. Such advanced instrumentation and avionics allows maximum utilization of the airplane in bad weather. (Beechcraft Photo)**

Due to the fact that the small bar magnets in a plain magnetic compass attempt to keep themselves aligned with the Earth's lines of flux, and due to the fact that these bar magnets are mounted on a pivot so that they have free turning which involves some allowable tilting, the plain magnetic compass has many errors. First, it is only accurate in straight and level, unaccelerated flight. During a turn it will swing wildly, and not follow the aircraft's direction during the turn. Second, turbulence causes the plain magnetic compass to oscillate and swing. Third, acceleration and deceleration cause the compass to indicate a turn when the craft is actually flying straight and level. For these reasons, one should be careful in reading the magnetic compass, and careful in setting a directional gyro to the magnetic compass heading. Of course, the gyro-stabilized flux gate compass has none of these errors.

The gyros in most flux-gate compasses will operate properly in most attitudes of flight except when the aircraft exceeds 85° of pitch or bank. At this point they "tumble," and the compass will spin wildly until reset. For this reason, if your aircraft has a gyro-stabilized compass, and you intend to do aerobatics with it, do not turn the compass on. Leave it off so that the gyro will not be tumbled during extreme maneuvers.

**2. Airspeed Indicator**

The first airspeed indicator was a pendulum which the slipstream pushed aft. The further aft the pendulum was pushed, the faster the aircraft was going. As early as 1913 a fellow over in Europe decided that a better airspeed indication would be obtained by comparing static pressure of the air to the ram air pressure developed by the aircraft's forward motion. This principle is still being used today. A ram air tube projects into the slipstream to measure the pressure of air developed by the aircraft's forward momentum. This ram air tube is called a pitot tube (French word, correctly pronounced pee-toe). It is connected to a tiny metal bellows or diaphragm. As the airspeed increases, the pressure in the pitot tube increases, expanding the bellows. A needle is mechanically geared to the bellows, and rotates on a dial calibrated in miles per hour or knots.

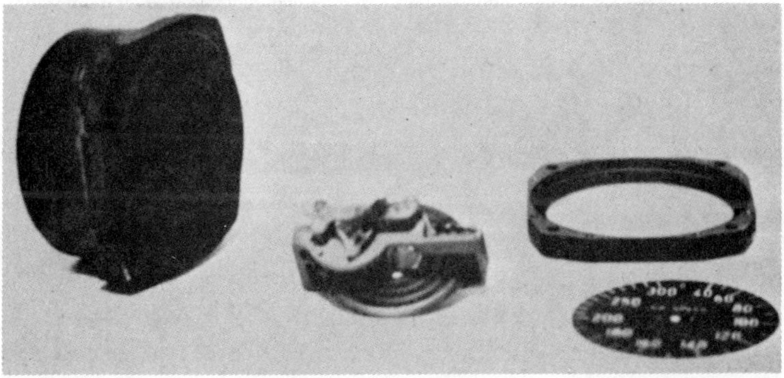

**Disassembled airspeed indicator. Assembly in center of photo is bellows which expands to give airspeed indication.**

The air around the bellows is vented to atmospheric, or static, pressure. Static ports are provided on the pitot tube itself or mounted on the aircraft skin. Some pitot tubes are designed with a line for measuring static pressure which, of course, is entirely separate from the pitot tube itself. In any case, the static pressure is measured at some point on the aircraft (or adjacent to the pitot tube), and vented to the airspeed indicator case around the bellows. Thus, the bellows is measuring the difference between static and pitot (ram) air pressure. This difference is displayed on the instrument dial in either miles per hour or knots. Some airspeed indicators show both miles per hour and knots.

**Older model airspeed indicator dial face. Newer airspeed indicators combine miles per hour and knots on the scale. This older one is in miles per hour.**

### 3. Altimeter

The earth's atmosphere is such that static pressure decreases as altitude increases. The pressure of the air at sea level is about 15 pounds per square inch, or about enough pressure to raise a column of mercury 29.92″. Sea-level pressure is thus expressed in inches of mercury. As one ascends in the atmosphere, the pressure lowers, and the column of mercury drops since less pressure is available to hold it up. The pressure then would be expressed in a smaller number of inches of mercury.

The rate of change of atmospheric pressure is approximately one inch of mercury (1″ Hg) per thousand feet of altitude. An aneroid barometer is an instrument in which an aneroid, or diaphragm, expands or contracts as atmospheric pressure increases or decreases. The aneroid (diaphragm) is mechanically connected to a needle on a dial to indicate pressure in inches of mercury. (In Europe, air pressure is measured in millibars, and aircraft altimeters are so calibrated.) If we take an aneroid barometer, and change its dial face to show feet, we will have an altimeter. The altimeter consists of a diaphragm which is sealed with 29.92″ Hg pressure (sea-level standard pressure) inside it. The instrument case around the diaphragm is sealed, and connected to static, or atmospheric pressure (static pressure *is* atmospheric pressure). Thus, the altimeter compares the static pressure of the atmosphere with the 29.92″ Hg pressure sealed in the diaphragm. The diaphragm is mechanically geared to needles which point to the difference between 29.92″ Hg and static pressure as expressed in feet above Mean Sea Level (MSL).

This arrangement works fine as long as the pressure at sea level remains at 29.92″ Hg—which it seldom does! When the pressure at sea level varies above and below 29.92″ Hg, the altimeter's altitude will vary above and below the correct altitude to the same degree that the pressure at sea level is varying. How is one to compensate for this? Fortunately, altimeter manufacturers provide an adjustment for just this sort of thing. A knob is provided on the front of the altimeter which may be adjusted to the proper sea-level pressure as shown in a small window on the altimeter dial face. The adjustment squeezes or expands the bellows to simulate lower or higher than 29.92″ Hg sea-level pressure. This causes the altimeter to indicate the proper MSL (Mean Sea Level) altitude when the adjustment is made for non-standard sea-level pressure.

Actual sea-level pressure can be obtained by calling the tower and the altimeter knob adjusted so that the pressure the tower gives you shows in the window. The altimeter should now indicate the airport's elevation above sea level, plus or minus 75 feet. If the altimeter is off more than 75 feet it should not be used for actual IFR flight. It needs adjustment, and possibly overhaul.

**4. Vertical Speed Indicator**

As we mentioned before, when one climbs or descends in the atmosphere, the pressure of the atmosphere decreases or increases. The vertical speed indicator measures this rate of change of atmospheric pressure by comparing movement of air through a vented hole to movement of air through a calibrated hole. The vented hole is connected to a bellows or diaphragm. A needle is geared to this bellows, and moves on a dial face calibrated in hundreds of feet of climb or descent. The instrument case is vented, via a calibrated "leak," to static atmospheric pressure just as the diaphragm is freely vented to static atmospheric pressure.

Thus, as the airplane climbs or descends there is a difference in pressure between the vertical speed instrument case and the bellows, caused by the fact that the air can change rapidly within the bellows but only slowly outside of it. This difference in pressure is displayed as a rate of climb or descent expressed in feet per minute.

There is a lag between the aircraft's actual performance, and the instrument indication when the aircraft is rapidly changed from straight and level to a climb or descent. Thus, the needle on the vertical speed instrument lags behind what the aircraft is actually doing when a rapid pitch change is made. When the pitch change is slow, the vertical speed will tend to indicate properly with little or no lag.

**5. Turn and Slip Indicator**

The Turn and Slip Indicator is really two instruments in one. The turn needle is gyro-operated while the slip indicator is a carpenter's bubble level. It is used to indicate the quality of a turn. The bubble, or '"ball", will be centered in straight and level flight and in a properly coordinated turn.

The first aircraft instrument to utilize a gyro was the turn indicator. The turn indicator uses a gyro's principle of precession to indicate the amount and direction of an aircraft's turn. Because it shows the amount as well as direction, the turn indicator is sometimes referred to as a "rate gyro." As such it is used in gyro-stabilized compass systems to cut off the magnetic signal during turns.

The gyro of a turn indicator is mounted so that its spin axis is in the horizontal plane, and the gyro itself is arranged fore and aft, i.e. the plane of rotation is a vertical plane along

World War II German Turn and Slip Indicator is electrically driven, similar to instruments manufactured in this country.

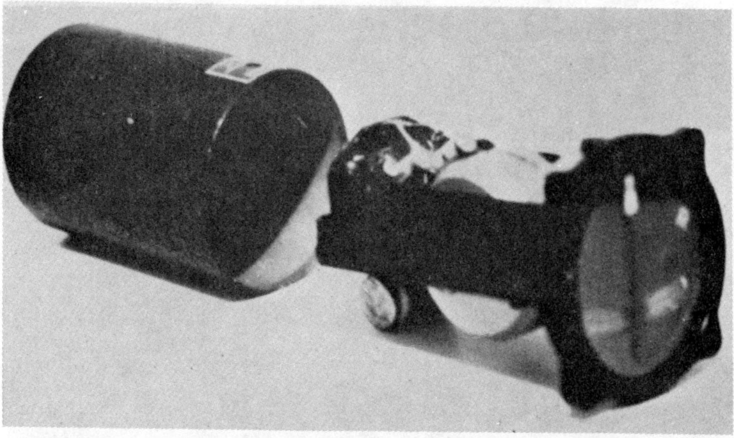

Electric Turn and Slip indicator with instrument case cover removed. Gyro rotor is armature of electric motor. Gyro instruments can be either electrically or pneumatically (vacuum) operated.

the aircraft's centerline. Precession causes a gyro to move in the same direction as the applied force, but 90° away in the direction of the gyro's spin. The gyro of a turn indicator is mounted so that whenever the aircraft yaws or turns the gyro precesses. This precession is transmitted to the turn needle via a mechanical linkage which causes the turn needle to deflect and show the direction and rate of the aircraft's turn or yaw.

**6. Gyro-Stabilized Heading Indicator**

As we have already seen, it is possible to use a gyro's principle of rigidity in space to stabilize the indications of a magnetic flux-gate compass. Before the development of the gyro-stabilized flux-gate compass, a gyro was used simply as a heading indicator. It had no magnetic reference, and had to be set with the magnetic compass. It could then be used to indicate magnetic heading, but it was sort of "second-hand" information. The gyro had to be reset frequently with the magnetic compass in order to be entirely accurate.

**Vacuum-driven Directional Gyro shown here has a ball-bank indicator on the dial face. Not all D.G.s have this feature. Knob on bottom of dial face is used to set azimuth card to agree with magnetic compass.**

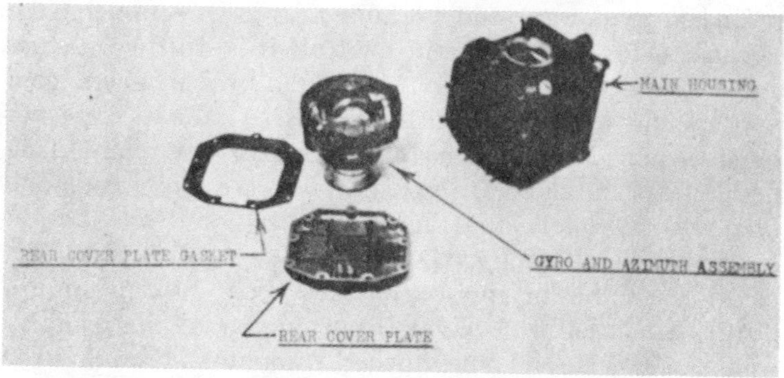


Disassembled vacuum driven D.G. azimuth card (or ring) can be seen on Gyro and azimuth assembly. Gyro is small wheel in center of Gyro and azimuth assembly. Notches are cut in the rim of gyro wheel so air pressure can spin gyro. Engine-driven vacuum pump or externally mounted venturi tube suck air through gyro housing where small jets direct it against notches in gyro wheel causing it to spin.

D.G. with rear cover plate removed clearly shows azimuth card (ring).

Such a gyro was (and is) known as a Directional Gyro. Because of its simplicity and low cost it is still in wide use, and will remain in wide use. Gyro-stabilized magnetic compass systems are too expensive for use on the average personal owner aircraft. A magnetic compass (or remote magnetic compass), and a Directional Gyro are more commonly used. Because of friction in the bearings, and other causes of precession, the D. G. (Directional Gyro) should be compared with the magnetic compass every fifteen minutes during an actual IFR flight. Remember that the magnetic compass is accurate only during straight and level, unaccelerated flight!

**7. Artificial Horizon**

The ultimate precision instrument for instrument flight is very obviously an instrument that will display the aircraft's attitude exactly as it is with respect to the earth's horizon.

**German Turn-Horizon is actually an artificial horizon and turn and slip indicator in one instrument.**

The Artificial Horizon is such an instrument. Using a gyro's principle of rigidity in space, the artificial horizon has an instrument face showing a tiny airplane superimposed on a horizon bar. The horizon bar is connected to the gyro in such a manner that it will show pitch and bank changes just as the actual aircraft is making them. In addition, a bank indicator is provided to show the exact angle of bank.

**German Turn-Horizon with case cover removed. Artificial Horizon gyro is in the center of instrument with turn gyro mounted vertically at the rear. Both gyros are electrically operated.**

The gyro in an artificial horizon takes several minutes to come up to operating speed and stabilize. When it does, the little airplane should be adjusted to show the proper straight and level flight attitude; all flight corrections will then be easier, and in the proper direction. The little airplane on an artificial horizon is movable up and down, and should be adjusted prior to flight and not changed until the aircraft reaches straight and level cruising flight. It then can be adjusted to show straight and level. Thereafter it should not be moved.

The artificial horizon, like all gyro instruments, has limits of operation. Its gyro will probably tumble if 55° of pitch is attained, or if the aircraft is banked 110° or more. Aerobatic maneuvers will damage the gyro, whether it is operating or not.

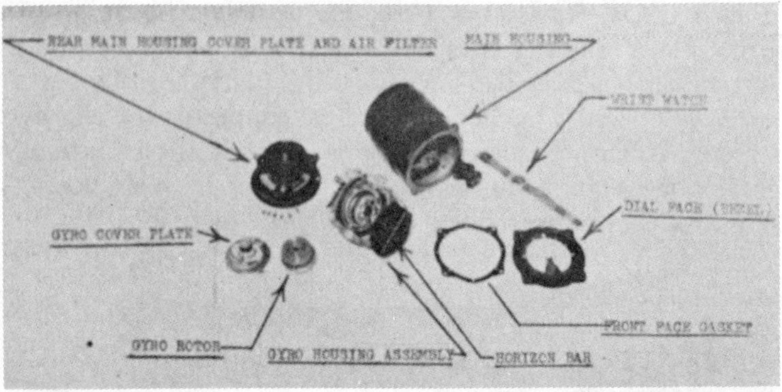


**Disassembled Sperry vacuum-driven Artificial Horizon. Wrist watch gives indication of size. Disassembled Gyro Housing shows Gyro Rotor which is very similar to gyro used in vacuum-driven D.G. Sperry vacuum Artificial Horizons were standard of the industry for over two decades, are being replaced by electric instruments and smaller vacuum-driven ones.**

## Summary of Flight Instruments

Flight instruments can be classified according to three types:

Magnetic heading indicators
Pitot-static system instruments
Gyro-stabilized instruments

Magnetic heading indicators range from the plain magnetic compass to gyro-stabilized compass systems. The plain magnetic compass requires no external source of power, but remote compasses and gyro-stabilized compasses require both DC and AC power to operate.

Pitot-static system instruments require no electrical power except to heat the pitot tube to prevent ice accumulation which will block the pitot tube and render the airspeed indicator inoperative. Anyone contemplating actual IFR flight should have a heated pitot tube installed on his aircraft.

Gyro-stabilized instruments include the turn and slip indicator, directional gyro (and gyro-stabilized compasses), and artificial horizon indicators. The gyros themselves may be rotated by air pressure supplied from either a venturi tube, or engine-driven vacuum pump. They may also be rotated by electrical power which is normally AC for directional gyros and artificial horizons, and DC for the gyro in turn and slip indicators.

All gyros require several minutes to come up to normal operating speed after power is applied, or the engine started in the case of air-pressure operated instruments. Where a venturi tube is used for air-pressure operated gyros, the aircraft must be in flight several minutes to bring the gyros up to speed. Even when the venturi tube is mounted in the propeller slipstream, not enough vacuum is developed to rotate gyros to operating speed until several minutes after the aircraft is airborne. For this reason, when you have a venturi tube, it is not safe to take off when the ceiling and visibility are so low as to preclude several minutes of flight in VFR conditions. Also, ice will accumulate on the venturi in icing conditions, rendering the gyro instruments inoperative.

All aircraft instruments should be removed and sent to a reliable instrument repair station periodically. Any good instrument repair station can advise you as to the safe time interval between overhauls for each of the instruments in your aircraft. When you're flying in IFR conditions, your life depends on having good instruments in proper operating condition. Be sure yours are in first-class order!

## 3. Basic Instrument Flight Maneuvers

Basic flight maneuvers for both instrument and contact (VFR) flying are the same: 1) Straight and Level, 2) Climbs, 3) Descents, and 4) Turns.

In contact flying we establish an aircraft attitude with respect to the earth's horizon, and an engine power setting with the manifold pressure and tachometer. We then observe the airspeed, altimeter, and compass to ascertain that the performance of our aircraft is what we desire. If it isn't, then we crosscheck the earth's horizon, and the engine tachometer (and manifold pressure for aircraft with this instrument), and adjust them as needed to obtain the desired performance. In instrument flying we do exactly the same thing except that we use an artificial horizon instead of the earth's horizon, and we rely solely on the performance instruments.

In contact flying we sometimes use the "seat of our pants" to aid our performance instruments. When flying by instruments, we must completely disregard the "seat of our pants," and rely entirely on the instruments.

For instrument flight a hood or visor is used so that the student cannot see outside the cockpit. There are several reasons for using a hood or visor, and preventing the student from "cheating" by sneaking a glance outside is *not* one of them. The primary purpose for use of a hood or visor is to simulate, as accurately as possible, real instrument flight conditions. One may fly simulated (hooded) instruments for many years and think he's a real pro at instrument flying, but there's nothing quite like that first flight in real instrument conditions. The more realistically the condition has been simulated in training, the easier the first real instrument flight will be. For this reason, the full cloth hood over the rear seat of a tandem trainer is the best for instrument flight simulation. However, the hood or visor used will depend on the individual situation, the particular aircraft, and the flight instructor. In any case, the object is to simulate flight in real instrument conditions, i.e. within clouds with zero visibility.

One of the most important lessons to learn about instrument flying is that our touch and feel senses will not be useful to us under actual instrument conditions and may, indeed, be harmful. For these senses we substitute the flight instruments. It is a paramount rule of instrument flying that the pilot *must believe* the instruments rather than his sense of touch, feel, or balance.

There will be times during actual or simulated instrument flight when a pilot will feel the airplane is not doing what his flight instruments tell him it is doing. For example, he may feel that he is in a turn when the flight instruments tell him the airplane is straight and level. This feeling can be so strong that a pilot must consciously fight it, and remember to continue to believe the flight instruments. This condition is a form of *vertigo*. Another example of vertigo, and a much more dangerous one, is caused by the pilot diverting his attention too long from the instruments to look at a map or chart, then coming back to the instruments and finding himself completely lost as to the aircraft's attitude. Since it is possible to become thus disoriented under actual or simulated instrument flight conditions, a pilot must learn proper instrument flying techniques, and maintain a high degree of proficiency in their use.

We divide the instruments into three major groups—Control Instruments, Performance Instruments, and Navigation Instruments. In this chapter we will examine Control and Performance Instruments, leaving Navigation Instruments for a later chapter dealing with radio aids to navigation. We use Control Instruments to control the aircraft's attitude and Performance Instruments to observe the aircraft's performance and ascertain that it is what we desire it to be.

# Axes of rotation

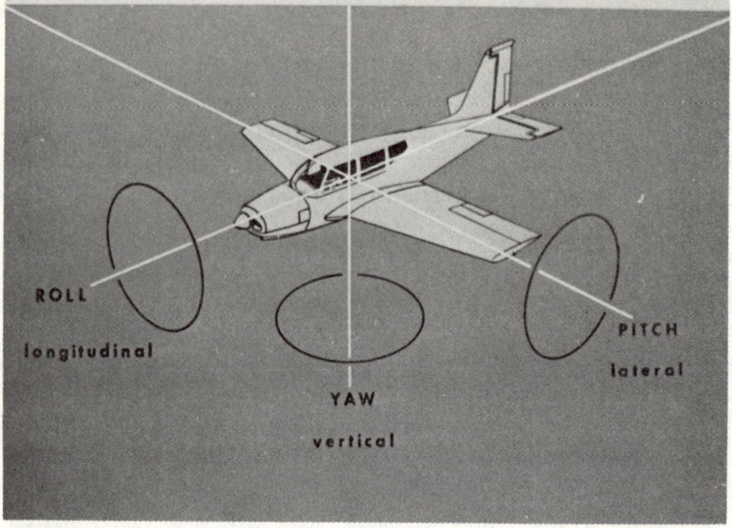


**Principal axes of an airplane. Pitch is controlled by the elevators and causes the aircraft's nose to move up or down. This is shown on the Artificial Horizon as movement of the miniature airplane relative to the horizon bar. Yaw is controlled by the rudder, and is shown jointly by the turn needle and ball-bank indicator. Bank is shown on the Artificial Horizon as tilting of the miniature airplane, and is controlled with the ailerons.**

We know that a given aircraft attitude plus a given engine power setting will result in a certain aircraft performance. Our Control Instruments are the artificial horizon and the engine power instruments: RPM and manifold pressure. We establish a given aircraft attitude on the artificial horizon, and set a given power setting on the RPM and manifold pressure. We know that what we have set should result in a certain aircraft performance. We then observe the Performance Instruments to determine if our attitude and power setting are giving the desired performance. If they are not, then we go back to the Control Instruments and adjust either the attitude or power setting. This procedure of observing the instruments is called "cross-checking," and all other things being equal, the faster a pilot's cross-check, the smoother he flys an aircraft under instrument conditions. This is also the reason why instrument training will improve the smoothness with which you control the airplane when flying contact.

### Straight and Level Flight

By definition, straight and level flying occurs when the aircraft flys a straight path through the air at a constant altitude and at a constant airspeed. We watch the D. G. (Directional Gyro) to determine whether or not we are flying a straight path through the air by observing whether or not the D. G. remains on the desired aircraft heading. The altimeter is watched to ascertain that we are flying a constant altitude, and the airspeed indicator is used as a reference instrument for constant airspeed.

Straight and level flight is normally accomplished at cruising airspeed, so our engine power will be normal cruise power. The engine power will determine our airspeed for any given flight attitude. The wing develops lift to support the airplane. At a slow airspeed, the wing's angle of attack must be higher than at a high airspeed. This means we must hold a higher (nose higher) pitch attitude for slow-speed flight than for high-speed flight. Normal cruise is accepted as the "level" pitch attitude, and our artificial horizon's miniature airplane should be adjusted to superimpose on the horizon bar at normal cruise airspeed. When the airspeed is slower than normal cruise airspeed, the miniature airplane will be slightly above the horizon bar indicating a nose-high pitch attitude. When the airspeed is higher than normal cruise airspeed, the miniature airplane will be slightly below the horizon bar indicating a nose-low pitch attitude.

To control pitch during straight and level flight, we establish the proper pitch attitude on the artificial horizon, and adjust the throttle for the proper RPM and manifold pressure. Then, we cross-check the altimeter to ascertain that we are maintaining the proper altitude. If the altimeter is below the proper altitude, we go to the artificial horizon and raise the nose to cause the airplane to climb slowly back to the proper altitude. If the altimeter indicates we are flying above our desired altitude, we lower the nose on the artificial horizon to cause the airplane to slowly descend back to the desired altitude.

Pitch changes made on the artificial horizon are made in increments of one horizon bar width, or fractions thereof.

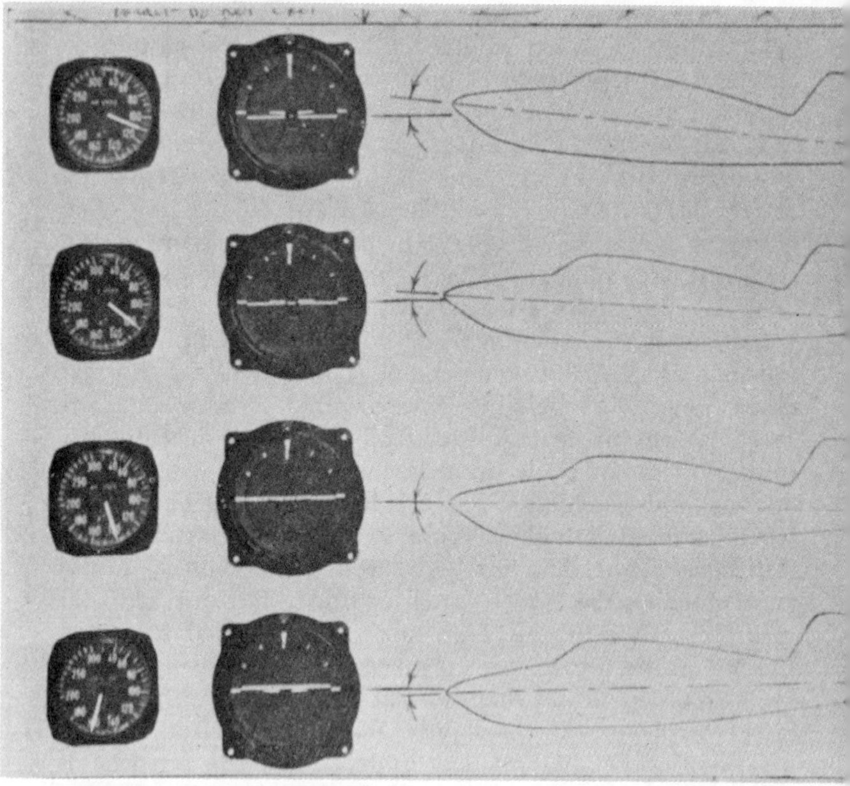

**Pitch attitude changes with airspeed when maintaining a constant altitude on the altimeter. Airplane shown above has a normal cruise speed of 140 mph (122 knots). Notice that, at slower airspeeds (120 and 110 mph), the nose must be raised on Artificial Horizon in order to maintain same altitude. At a higher airspeed, less lift is required, so nose will ride lower on horizon.**

For small pitch changes, ½ or even ¼ bar width change in pitch attitude may be sufficient. For example, if the altimeter indicates we are 50 feet low, raising the miniature airplane a distance of ¼ bar width on the artificial horizon should be sufficient to cause a change in altitude at a rate consistent with good aircraft control. The important point is to avoid overcontrolling. Overcontrolling results when the control instruments are adjusted too much for the desired change in the performance instruments. Be smooth, and make small corrections.

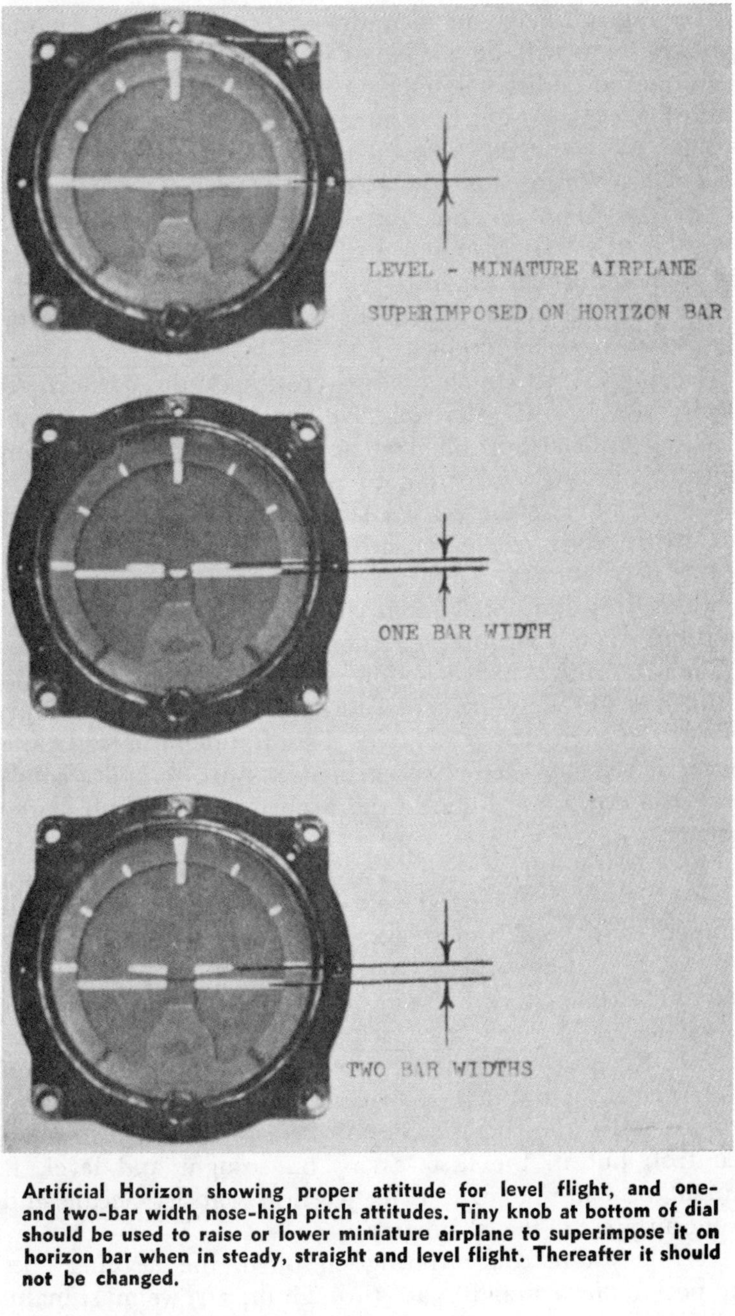


**Artificial Horizon showing proper attitude for level flight, and one- and two-bar width nose-high pitch attitudes. Tiny knob at bottom of dial should be used to raise or lower miniature airplane to superimpose it on horizon bar when in steady, straight and level flight. Thereafter it should not be changed.**

During a change in altitude at constant cruise engine power, there will be a change in airspeed which will necessitate an additional pitch change on the artificial horizon. For example, let us assume we are 100 feet low on altitude. We raise the nose on the artificial horizon bar ½ bar width. This causes the airspeed to decrease which may cause the climb to stop short of our desired altitude. We must then go back to the artificial horizon and raise the nose an additional amount—say an additional ¼ bar width. This is the time at which the vertical velocity indicator begins to have importance.

During a small pitch change, crosscheck the vertical velocity indicator. If, after you have established a pitch change on the artificial horizon, you note the vertical velocity indicator is slowly returning to zero, you may assume it is doing so because the pitch change has caused the airspeed to change which means you need an additional pitch change. Crosscheck both the altimeter and airspeed indicator before making this additional pitch change. The vertical velocity indicator is not an instant response instrument at all times. When the pitch change is slow, the vertical velocity needle will tend to follow the actual rate of climb or descent. But, when the pitch change is rapid, there is a definite lag in the vertical velocity needle, and it may require several seconds to settle down and indicate the proper rate of climb or descent.

Be careful during early pitch control lessons that you don't "chase" the vertical velocity needle. It is a common student error, and almost every pilot will have a tendency to do it. However, place your primary pitch performance on the altimeter, and use it until you become proficient enough to add the vertical velocity indicator to your crosscheck. Keep in mind that the vertical needle will lag when you make rapid pitch changes. Once you become accustomed to its idiosyncrasies, it will be a valuable aid to pitch control; but in the first lessons on straight and level, it might be better to ignore the vertical velocity needle and concentrate on the altimeter.

While you're concentrating on pitch, the heading goes to pot! To fly a straight path through the air, we must maintain a constant D. G. (Directional Gyro) heading. As long as the wings are level, and there is no yaw caused from uncorrected torque effect, or having the ball out of center in the turn and slip indicator, the D. G. will maintain a constant heading. To maintain a constant D. G. heading, we keep the wings level.

If the D. G. turns while the wings are level, our rudder must not be properly trimmed. If the D. G. turns while the wings are level, the ball in the turn and slip indicator will be out of center, and the turn needle will be deflected in the direction of turn. Therefore, if the wings are level on the artificial horizon, and the D. G. is still turning, the problem is incorrect rudder trim.

Apply rudder pressure to stop the D. G. from turning and, if the aircraft is equipped with a rudder trimming tab, adjust the tab to relieve the rudder pressure. The ball in the turn and slip indicator indicates the quality of a turn, and indicates the quality of a straight and level flight so far as rudder trim is concerned. If the rudder is not trimmed properly the aircraft will be slipping or skidding and the ball will be displaced from center, whether in straight and level flight or in a turn. Improper rudder trim is the result of uncorrected (by rudder) torque effect, or the rudder trim tab being out of proper adjustment.

Whatever the cause, the solution is to apply rudder on the side on which the ball is out of center. For example, if you find the wings are level on the artificial horizon, and the ball is off to the right, apply right rudder until the ball is centered. Remember that moving the rudder will have a tendency to cause the aircraft to roll, so quickly crosscheck the artificial horizon to see that the wings remain level while the ball is being centered. Straight flight requires that the wings be level, with the ball-bank indicator centered and the D. G. (or gyro stabilized magnetic compass) maintaining a constant heading.

An aircraft will turn any time the wings are banked. The rate of turn depends on the amount of bank, and is in direct proportion to the amount of bank. A small amount of bank gives a small rate of turn. Increasing the bank increases the rate of turn. When attempting to maintain straight and level flight, corrections back to the desired D. G. heading should be smooth, and small in amount. A good rule is to bank the same number of degrees the D. G. is off the desired heading. For example, if you are trying to maintain a heading of 090°, and you note the D. G. is on 080°, you are 10° off the desired heading. Using coordinated aileron and rudder pressures, roll the miniature airplane on the artificial horizon into a 10° right bank, then crosscheck the D. G. As it approaches the desired heading, use coordinated aileron and rudder pressure to roll the wings level on the artificial horizon. As with pitch corrections, the object is to be smooth, and do not overcontrol.

Torque effect causes the nose of an airplane to swing to the left as engine power is added (or nose is raised), and swing to the right when engine power is reduced (or nose is lowered). While small changes in torque are unconsciously corrected for in VFR flight, and never bother the student pilot, such changes on instruments tend to wreak havoc with aircraft control. The main reason is because control of the aircraft on instruments is generally more precise than when flying VFR, and small torque changes are more readily detected.

If, while attempting to maintain straight and level flight, the aircraft gets 100 feet low on altitude, and the pilot raises the nose to gain altitude at the rate of 100 feet per minute, the airspeed will decrease slightly, and the ball in the turn and slip indicator will move ever so slightly to the right. This is because airspeed has decreased for a given amount of engine torque. Similarly, increasing engine power always causes the nose of the aircraft to swing to the left. If, while making this small altitude change, the pilot observes the D. G. he will find it has turned a few degrees to the left, Why? Torque. The conclusion then, is that during small altitude changes, the pilot must observe the ball in the turn and slip indicator and the D. G., to be sure he is not turning. He can anticipate that raising the nose will cause the ball in the turn and slip to swing to the right, and that a tiny amount of right rudder will be necessary to center it, and prevent a turn due to torque. *The reverse will be true during a descent.*

Straight and level flight requires a good crosscheck of all the flight instruments and engine power instruments. The first lessons in straight and level should consider only pitch control, then bank control, then the two should be put together. Control movements should be smooth, but positive. Use all available trim controls. Establish the proper attitude, allow the performance instruments to settle down, then relieve control pressures with whatever trim controls are available.

**Turns**

An airplane is turned by tilting the lift of its wings to one side or the other. Tilting the wings' lift causes it to pull in the direction it is tilted, or *banked.* The more it is banked, the greater it trys to pull to that side, and the faster the airplane will turn. Because some of the lift is being used to turn the airplane, a portion of the lift being used to maintain altitude is lost. The more the airplane is banked, the more of its vertical lift is lost. If not corrected, the airplane will begin to descend. It can be corrected by raising the nose on the artificial horizon to maintain level flight during a turn. The greater the angle of bank, the more the miniature airplane will have to be raised on the horizon bar to maintain a constant altitude during a turn. It is also possible to prevent a descent during a turn by adding power to increase airspeed, but this is not normally done except in very steep turns. The maximum amount of bank while flying instruments should be about 30°. This will not normally require any addition of power, but will require some raising of the nose during the turn.

We enter a turn by rolling in on the artificial horizon. Smoothly coordinate aileron and rudder pressure to cause the aircraft to roll. Observe the bank indice on the artificial horizon, and as it approaches the desired angle of bank, relax the aileron and rudder pressure. Crosscheck the altimeter for any necessary pitch changes, and make them just as you did in straight and level flight. Crosscheck the turn and slip indicator, and center the ball using the same techniques developed in straight and level flight practice. Once a turn is established, the crosscheck is the same as in straight and level flight, the only difference being that the aircraft is banked, and is therefore turning.

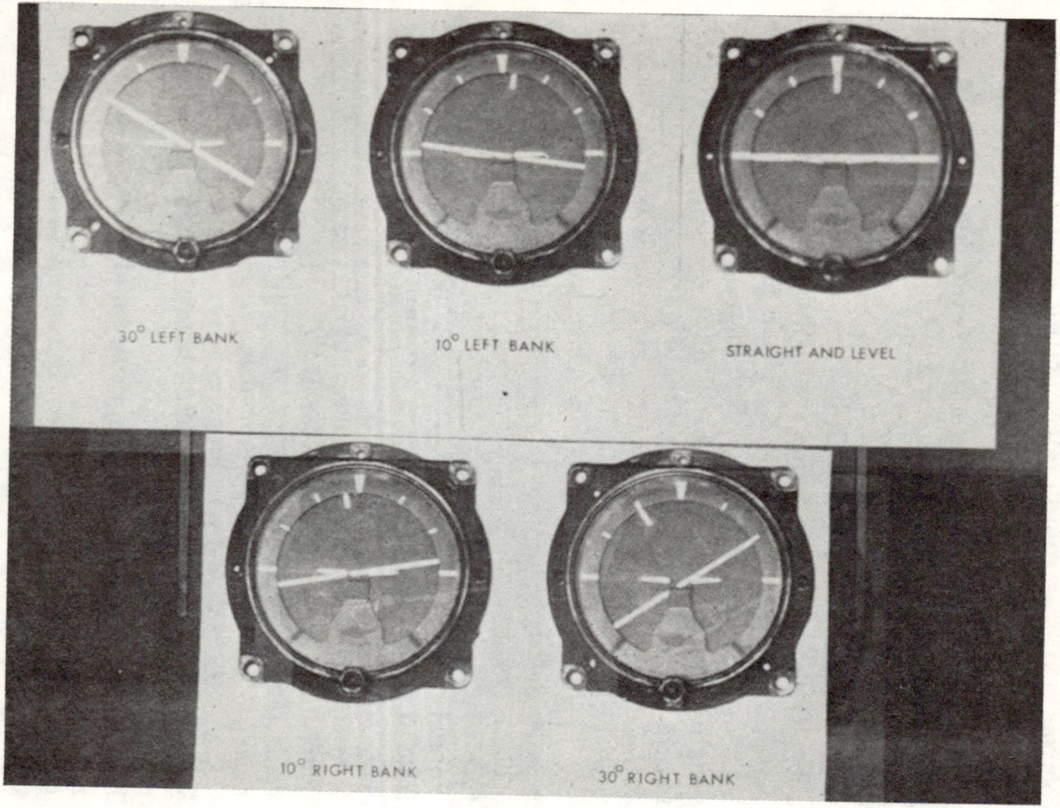


Bank attitudes on Artificial Horizon for straight and level, and 10° and 30° of left and right bank.

While rolling into a turn you can expect to have to raise the nose to prevent losing altitude, but don't anticipate this, or you'll gain altitude! Crosscheck the vertical velocity and altimeter, and when you see a need for a pitch change, be smooth and positive in making it.

Rolling out of a turn is as simple as rolling in. Coordinate aileron and rudder pressure to roll the miniature aircraft to a wings-level flight attitude. Crosscheck the altimeter for any necessary pitch changes. Again, don't anticipate any pitch changes. Although you can expect to have to lower the nose when rolling out of a turn, if you anticipate it, you are bound to overcontrol. Wait until the altimeter and/or vertical velocity indicator indicate a need for a pitch change. Then make it smoothly and positively.

**Straight Climbs**

The transition from straight and level cruising flight to a straight climb involves a change in pitch, power, and torque correction. We simultaneously raise the nose on the artificial horizon and add climb power on the engine by referring to the engine instruments. Adding power causes the nose to swing left, so we must crosscheck the ball in the turn and slip indicator, and add sufficient right rudder to keep the ball centered. As the airspeed decreases to climb airspeed, torque effect will be even more pronounced. Continue holding the wings level on the artificial horizon and crosscheck the ball in the turn and slip indicator. As the airspeed decreases to climb airspeed, increasing right rudder will be needed to keep the ball centered.

As the airspeed approaches climb airspeed, adjust the pitch attitude on the artificial horizon to maintain the proper climb airspeed for your aircraft. If the airspeed is too low, lower the nose by referring to the artificial horizon. If the airspeed is too high, raise the miniature airplane on the artificial horizon. When the climb airspeed has been definitely established, trim the airplane.

During a normal climb, constant reference must be made to the airspeed indicator in order to maintain the proper climb airspeed. When the airspeed is low, you are climbing too steeply. Lower the nose on the artificial horizon. If the airspeed is low by five knots, try using a 1/4 bar width correction. Give the airspeed time to change, and if this correction isn't enough, make an additional 1/4 bar width correction. If the airspeed is low by ten knots, try using a 1/2 bar width correction. Give the airspeed time to change before making further changes on the artificial horizon.

When the airspeed is high, you're not climbing steeply enough, and you must raise the nose on the artificial horizon. Until you determine the proper pitch changes for your aircraft, use 1/4 bar width for five knots and 1/2 bar width for ten knots. Hold the change constant on the artificial horizon until the airspeed has had time to change, then make any additional pitch changes as indicated by the airspeed indicator.

Maintain bank and heading control during climbs just as you did in straight and level flight and turns. A common student error is to forget heading and bank during climbs. Remember that good instrument flying requires that the airplane be controlled on all axes, which means *all* the flight instruments must be crosschecked *all* the time!

To level off from a climb, lower the nose on the artificial horizon as you approach the desired altitude. Adjust the pitch attitude on the artificial horizon as necessary, to maintain a constant altitude on the altimeter during the level off. As the airspeed increases, the nose will have a tendency to swing to the right due to torque effect, and rudder must be used to correct it. As the airspeed approaches normal cruise, smoothly reduce power to cruise power, and begin a normal straight and level crosscheck. Trim the aircraft.

**Straight Descents**

A descent is, of course, the exact reverse of a climb. To enter a descent, retard engine power to descent power. Maintain level flight by crosschecking the altimeter, and making necessary pitch changes on the artificial horizon to maintain a constant altitude until the airspeed approaches normal descent airspeed. Then, lower the nose on the artificial horizon to maintain descent airspeed. Pitch control during descent is exactly the same as during a climb. Control reaction time will be slower due to the slower airspeed, and due to reduced propeller wash over the empennage, but the amount of pitch change required to effect a given airspeed change will remain about the same.

As engine power is reduced from cruise power, the nose will have a tendency to swing to the right due to torque effect. Corrections should be made just as in climbs and in straight and level flight. The only difference is that, during a descent, propeller wash over the empennage is reduced, and control pressures are lighter due to the slower airspeed. More control movement may be required to effect the same results, but in any case, use whatever control is necessary to keep the wings level, and the ball centered.

Be mindful of bank and heading control, and make corrections early. Be smooth and positive, and do not get in such a hurry that you overcontrol.

As you approach the desired level off altitude, smoothly apply engine power to normal cruise. As the airspeed increases, adjust the pitch attitude on the artificial horizon to maintain the desired level off altitude on the altimeter. When steady, straight and level flight has been attained, trim the aircraft.

## 4. Advanced Instrument Flight Maneuvers

Before attempting advanced flight maneuvers it is absolutely essential for the pilot to master the basic maneuvers of straight and level, turns, climbs, and descents. *All further instrument maneuvers are based on, and presume proficiency in, the basic instrument flight maneuvers.* It is highly desirable, at this stage, that the pilot understand trim and how it should be accomplished. The basic technique for proper trimming for any maneuver, is to establish a steady flight condition, hold it, then remove control pressures with whatever trim tabs are available. Proficiency in trimming the airplane is not likely to come for many, many hours, but at least at this stage, the pilot should understand how it is done, or should be done.

### Airspeed Change in Level Flight

Changing airspeed in level flight gives practice in trim ming, engine power changes, and basic aircraft control. Airspeed changes should be from one definite airspeed to another definite airspeed such as from normal cruise down to holding airspeed. In aircraft where normal cruise is the holding airspeed, the speed change should be from normal cruise down to normal climb airspeed in straight and level flight. During practice airspeed changes, the new airspeed should be held for several minutes before returning to the original airspeed. Emphasis should be on smoothness and proper trimming.

To reduce airspeed from normal cruise to holding airspeed (or normal climb airspeed, whichever is applicable to your aircraft), smoothly reduce power to a power setting slightly below the power setting required to hold the lower airspeed. For example, if you have an aircraft that normally cruises at 23 inches manifold pressure, and you have found that 19 inches manifold pressure will hold climb airspeed in level flight, then reduce power to approximately 15-16 inches manifold pressure. Remember that torque will change, and that, as the airspeed decreases, the pitch attitude must be raised on the artificial horizon to maintain altitude.

As the airspeed approaches the lower airspeed, add power to hold the lower airspeed. Continue to crosscheck the altimeter and heading indicator (D. G.) to be sure you are maintaining straight and level flight. Make pitch changes on the artificial horizon as necessary, and check the ball in the turn and slip indicator to be sure it is centered. When the lower airspeed is being held steady, trim the aircraft.

To increase airspeed from holding to cruise airspeed, add power smoothly to a power setting slightly in excess of normal cruise power. There will be an immediate torque change, as is always true when changing power, and the turn and slip indicator must be checked, and the necessary rudder applied to keep the ball centered.

As the airspeed increases it will be necessary to lower the nose on the artificial horizon in order to maintain a constant altitude on the altimeter. As the airspeed approaches normal cruise, reduce power to normal cruise, and once a steady flight condition is attained, trim the aircraft. As your trimming proficiency increases, you will find it helpful to trim during a change in airspeed or other maneuver. but care must be taken not to overtrim.

**Airspeed Changes in Turns**

Changing airspeed once a steady turn is established is no more difficult than changing airspeed in level flight, and the procedures are the same. Some additional attention is necessary to maintain the proper bank, but otherwise, the techniques and procedures are the same as for changing airspeed in level flight.

Changing airspeed while entering a turn, or rolling out of a turn, is quite a different matter. Although not a difficult maneuver, it does require good concentration, planning, and a fast instrument crosscheck. There are few instrument maneuvers that will develop one's proficiency faster than practicing changing airspeed while entering or rolling out of a turn.

There are four variations of changing airspeed while entering or rolling out of a turn:

1. Entering a turn and decreasing airspeed
2. Entering a turn and increasing airspeed
3. Decreasing airspeed while rolling out
4. Increasing airspeed while rolling out

In all four cases, the engine power is changed as control pressures are applied to change the aircraft's bank.

Entering a turn while simultaneously decreasing airspeed means that decreasing airspeed will necessitate a higher pitch attitude on the artificial horizon, and entering a turn will also require a nose-high attitude. Torque is also changing so attention must be paid to the ball in the turn and slip indicator throughout the maneuver. The technique is to apply smooth control pressures to enter the turn meanwhile smoothly reducing engine power with the throttle. Establish the bank attitude on the artificial horizon, then rapidly crosscheck the altimeter for any necessary pitch change. Check the ball for any change in rudder pressures, and crosscheck airspeed and engine instruments for any needed power change. The vertical velocity indicator is a good "trend" instrument to indicate when a pitch change may be impending. The altimeter and vertical velocity instruments should both be checked for any pitch changes, but remember that the vertical velocity indicator has some lag. Any time you observe the vertical velocity needle moving, it is worthless as a positive pitch instrument, but highly valuable as an instrument that will tell you what is about to happen. For example, if you are entering a turn, and you notice the vertical velocity needle is moving rapidly upward, you may expect that you have raised the aircraft's nose too high as you entered the turn. But crosscheck the altimeter to see what it is doing before you make any change on the artificial horizon.

Rolling out of a turn and increasing airspeed requires a simultaneous application of control pressures to roll out and engine power to increase airspeed. Again, a rapid crosscheck of altimeter, airspeed, and the ball in the turn and slip indicator is necessary to properly control the aircraft and perform the maneuver smoothly.

**Walter Beech (front cockpit), experimenting with instrument flying technique in 1927, had gyro-controlled turn-bank indicator in his Travel Air 4000. Brice Goldsborough (rear cockpit) of Pioneer Instrument Company, tried out vertical-reading type gauges.**

When entering a turn and increasing airspeed, and when increasing airspeed while rolling out of a turn, the same procedures and techniques as just explained will apply. In all changes in airspeed when entering and rolling out of turns, it is essential to make a rapid crosscheck of all the performance instruments, and to make prompt and positive changes on the control instruments. But don't fall victim to the illusive habit of trying to "outguess" the aircraft, and make changes before they are required. Make changes on the artificial horizon and engine instruments only when the performance instruments indicate a change is necessary; and make the change smoothly and positively. Don't be a "Jerky Joe," pushing and shoving everything in sight! *Be smooth.*

**Rate Climbs and Descents**

*Now* we learn what the vertical velocity indicator is for. Rate climbs and descents are a necessary part of instrument flying, and are excellent practice in aircraft control with reference to instruments. During rate climbs and descents the object is to maintain a constant rate of climb or descent.

This requires careful coordination of pitch and power. We will assume that rate descents are performed at normal descent airspeed, and rate climbs at normal climb airspeed. Depending on the performance characteristics of your aircraft, use a rate of climb of 500 feet/minute, and a rate of descent of 500 feet/minute, or 300 feet/minute, whichever best suits your aircraft.

To enter a rate descent, we first maintain level flight and reduce power to descent power. Allow the aircraft to slow to the descent airspeed. As the airspeed approaches descent airspeed, lower the nose on the artificial horizon an amount that you think will give a 500 feet/minute (or 300 feet/minute, whichever you're using) rate of descent. Crosscheck the vertical velocity needle when it settles down. If the rate of descent is not correct, adjust the pitch attitude on the artificial horizon.

Make pitch changes on the artificial horizon to maintain the desired rate of descent. If the rate of descent is less than desired, go to the artificial horizon, and lower the nose a slight amount. If more than the desired rate of descent, raise the nose on the artificial horizon. Do not "chase" the vertical velocity needle by attempting to make pitch changes while the needle is moving.

Use the airspeed indicator as an indication of power required. If the airspeed is low, add power; if it is high, reduce power. Coordinate airspeed changes with pitch changes. For example, if the airspeed is high, and the rate of descent too fast, you may anticipate that no power change is necessary, but that a pitch change (raising the nose) will reduce both airspeed and rate of descent. In this manner, you may coordinate any combination of airspeed deviation and rate of descent deviation to attain the desired airspeed and rate of descent. A prerequisite is the ability to rapidly crosscheck the instruments, and to properly interpret them. Proper trimming is essential if any maneuver is to be smooth.

There are two methods of entering rate climbs. One way is to first reducc airspeed to climb speed while maintaining level flight; then apply climb power, and begin the climb. The other method is to raise the nose while at cruising airspeed, and allow the initial portion of the climb to dissipate the airspeed to normal climb airspeed. In the latter method, the initial portion of the climb is usually at a higher rate of climb than desired due to the higher entry airspeed. When the airspeed slows to normal climb airspeed, the pitch attitude is adjusted on the artificial horizon to the proper attitude for the climb.

When using the method of reducing airspeed, then beginning the climb, the initial portion of the climb is at the proper rate of climb; whereas, in the method of allowing the airspeed to slow from cruise to climb speed, the initial portion of the climb is at a higher rate, and the attitude on the artificial horizon must be higher than for a steady rate climb. It is then adjusted to an attitude that will maintain the desired rate of climb. As in rate descents, coordinate power and pitch changes to maintain the proper rate of climb, and the proper airspeed.

In both rate climbs and descents, a common error is to "chase" the vertical velocity needle. To avoid this, allow the needle to settle down before making pitch changes. Another common error is to forget torque. Remember that reducing engine power causes the nose of the aircraft to swing to the right—adding power causes the nose to swing to the left. During any maneuver (such as rate climbs and descents) where the heading is not being intentionally changed, it is essential to crosscheck the directional gyro for any undesirable change in heading, and correct it by changing the bank on the artificial horizon.

In making heading changes, the rule is to use an amount of bank equal to the number of degrees of turn that you're off the desired heading. In this manner you will not bank so much that you turn past the desired heading before you have a chance to crosscheck the D.G. Whenever it is necessary to change the bank of the airplane, remember to use coordinated rudder and aileron pressure just as you would during VFR flight.

Recovery from a rate climb or descent to level flight is the same as from straight climbs and descents. Be mindful of torque changes as airspeed and power are changed, and make your corrections smoothly, but positively.

**Precision Turns**

A precision turn may be defined as a turn to a precise, or specific, heading. Precision turns are essential in instrument flying. Techniques and procedures for entering and rolling out of turns are explained in Chapter Three, and will not be repeated here.

Turns to specific headings are more difficult using the conventional directional gyro than on 360° dial-type heading indicators. They are somewhat simplified if you remember that when the numbers on the D.G. are increasing, you're turning to the right; when the numbers are decreasing, you're turning to the left.

Entering a precision turn is the same as entry into any level turn (see Chapter Three). The precision part comes into play with the roll-out on a specific heading. One cannot wait until the D.G. is on the desired heading before beginning the roll-out. If you do, the D.G. will continue to turn in the roll-out, and will end up past the desired heading. It is necessary to "lead" the roll-out. By "leading" the roll-out, we mean you must begin to apply control pressures to roll the aircraft to level flight a certain number of degrees of turn prior to the desired heading. How much you must lead the roll-out depends on three factors: the aircraft's speed, angle of bank, and your technique. All of these can be boiled down to the one element you control—your technique. But first, we consider the angle of bank, and use it as a guide.

When you begin to practice precision turns, use the "one-third the angle of bank" rule. As your proficiency increases, and your technique develops, you can modify the one-third rule, but initially stick with it. The one-third the angle of bank rule says that you use one-third the angle of bank as the number of degrees at which to begin the roll-out. Thus, if you're in a 30° banked turn, lead the roll-out by 10° (⅓ of 30° is 10°). For example, if you're turning from 360° in a right turn to 090° using a 30° banked turn, you would begin the roll-out at 080° (⅓ of 30° is 10° and 10° left of 090° is 080°).

If, when using the one-third rule, you find you undershoot or overshoot the desired heading, adjust your start of roll-out point accordingly. This adjustment to the one-third rule is your technique. It does not come after a turn or two, but after many hours of making precision turns. Until you are satisfied that you know what your own lead should be, continue to use the one-third the angle of bank rule.

**Rate Turns**

An aircraft's rate of turn is displayed on the turn needle of the turn and slip indicator. There are two types of turn needles. One type is calibrated to show a 3°-per-second rate of turn with one needle width deflection from center. This amount of turn (3° per second) is known as a *standard rate turn*. The second type of turn indicator shows 1½°-per-second rate of turn for a single needle width deflection. This is a *half-standard rate turn,* and is usually installed on high-speed aircraft.

Any time the aircraft's heading is changing, the turn needle will be deflected. The amount of deflection depends on how fast the aircraft is turning. The faster the turn, the more the needle is deflected from center. The needle is calibrated, as we've just explained, to show a specific rate of turn for a specific deflection of the needle. We will assume that we're using a 3°-per-second turn needle, i.e., one needle width equals 3° per second of turn. Thus, a deflection of one-half the needle width would be 1½°-per-second rate of turn, and a deflection of two needle widths would be 6° per second rate of turn.

The turn needle is an instantaneous reading instrument. The instant the aircraft begins to turn—for any reason—the turn needle will deflect to show both the direction, and the rate of turn. For this reason, it is an excellent back-up instrument for maintaining straight flight. The primary instrument for straight flight is, of course, the heading indicator or D.G.

Normally we will never exceed a standard rate turn while flying on instruments. The angle of bank for a standard rate turn varies with the aircraft's true airspeed. At a low airspeed, a small amount of bank will give a standard rate turn. At very high airspeeds, a large amount of bank is necessary for a standard rate turn. Most of the aircraft we fly will require from approximately 10° of bank to approximately 20° of bank for a standard rate turn.

When the aircraft is equipped with a D.G. or gyro-stabilized compass system, rate turns have little value, and may be disregarded. However, rate turns are important when the only heading indicator is the standard magnetic compass. By performing a rate turn of definite duration, a specific number of degrees of turn may be accomplished accurately without reference to a heading indicator. One first must determine the number of degrees to be turned. For the standard rate turn (3° per second), divide the number of degrees to be turned by three, and the answer is the number of seconds the standard rate turn must be held. For example: We wish to turn 120° so we divide 120° by 3 which gives us 40. We will hold a standard rate turn for 40 seconds, and at the end of that time we should have turned 120°.

To perform a rate turn, begin the timing as control pressures are applied to roll into the turn. Establish the approximate angle of bank, and crosscheck the turn needle. If the rate of turn is too rapid, shallow the bank; if the rate of turn is too slow, steepen the bank. Adjust the bank on the artificial horizon to give one needle width deflection. Pitch and torque corrections are the same as for level turns. When the time has expired, begin rolling the wings level on the artificial horizon. Try to make the rate of roll-in and roll-out the same. If you are able to roll into the turn and roll out of it at the same rate, you will find that rate turns are extremely accurate. Their accuracy depends, of course, on how well you hold a constant single needle width deflection on the turn needle, whether you roll-in and roll-out of the turn at the same rate, and on the accuracy of the timing.

**Steep Turns**

Any turn where more than 30° of bank is used may be regarded as a steep turn. There is normally no reason to perform a steep turn while flying on instruments, but the practice of steep turns will increase the speed of your crosscheck, enhance the smoothness with which you control the aircraft, and give you practice changing engine power while in a turn. You should attempt to maintain a constant airspeed while making a steep turn. This will almost always require additional power due to the steep bank being used.

Rolling into a steep turn smoothly is essential if good

aircraft control is to be maintained. Rapid crosschecking of the altimeter and vertical velocity indicator is essential if the turn is to be level. Use of trim is recommended, but do not "chase" the trim. Trim as required to relieve excessive pressures. Use a specific amount of bank, and strive to hold it constant throughout the turn. Make the turn to a specific heading, but *never* try to make small turns—turn at least 180°. A turn of 360° is recommended for initial steep turn practice. Later they may be reduced to turns of 180°.

**Instrument Practice-Pattern "A"**

By now we are ready to combine our basic and advanced maneuvers into a combination maneuver designed to lead us into actual instrument navigation and approach practice. We call this combination practice maneuver Pattern "A" for lack of a better name. Prior to beginning the Pattern A, establish yourself on one of the cardinal points of the compass—090°, 180°, 270°, or 360°. Establish cruising flight and maintain the cardinal compass heading.

The entry point on the Pattern A is Point A. At this point, maintain the cardinal heading, and level, cruising flight. Begin timing a one minute leg. We suggest you use a watch or clock with a sweep second hand. Maintain the original heading, and cruising airspeed for this first one-minute leg. This leg ends at Point B. At Point B enter a standard rate turn to the right. After the turn is established (Point C), change airspeed to holding (or climb) airspeed and maintain level flight. Roll out of the turn at Point D which is 180° of turn.

At Point D begin timing another one-minute leg, and continue to maintain the original altitude, and holding (or climb) airspeed. After one minute, you are at Point E. At Point E, simultaneously begin a 45° left turn, and change airspeed to cruise. Maintain your altitude. Use a standard rate turn in all turns in Pattern A. Depending on the acceleration characteristics of your airplane, you may complete the 45° turn before arriving at cruise airspeed. If so, begin timing a fifteen-second leg as soon as the wings are level, and continue to accelerate to cruise airspeed. The fifteen-second leg ends at Point F. At Point F begin a 180° level turn to the right. After the turn is definitely established, change airspeed back to holding (or climb) airspeed. Roll out at Point

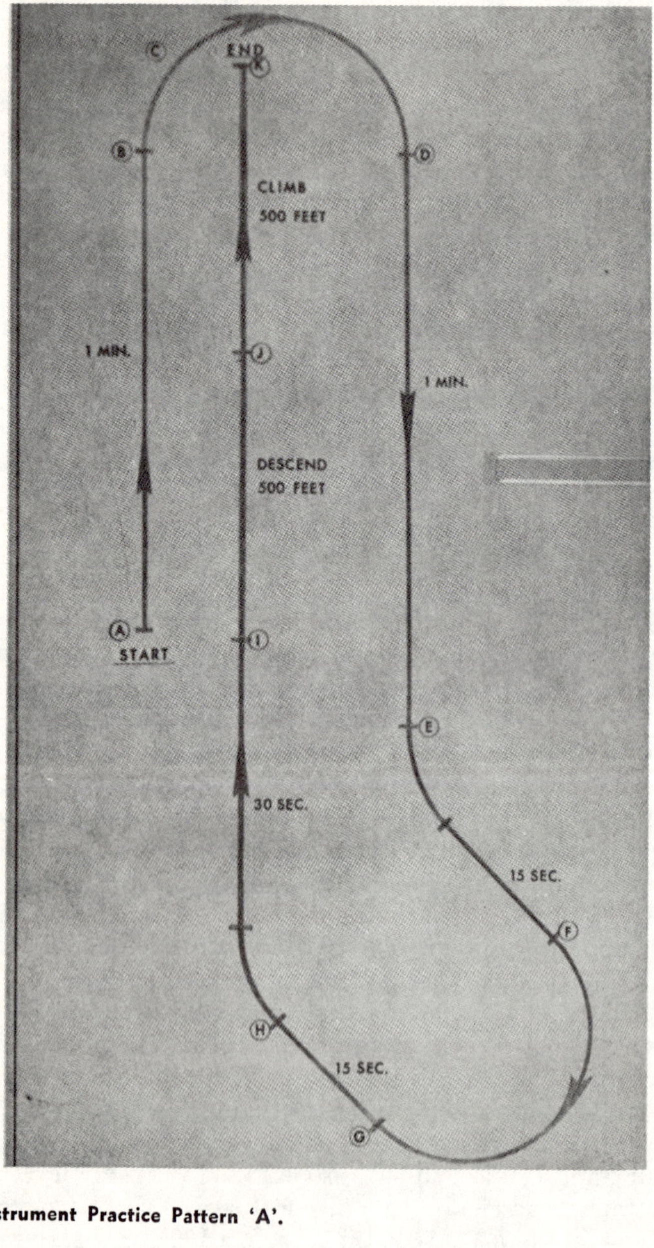


**Instrument Practice Pattern 'A'.**

G, and begin timing another fifteen-second leg. At Point H make a level 45° turn to the right and begin timing a thirty-second leg. After the timing has started, reduce airspeed to holding (or climb) airspeed, or the airspeed you have been using for rate descents. Continue to maintain level flight. At the end of the thirty seconds you are at Point I.

At Point I begin a rate descent (using 500 feet/minute or 300 feet/minute) for 500 feet. After descending 500 feet you are at Point J, and at this point, change from a rate descent to a rate climb using the same rate of climb you used in the descent. Climb for 500 feet. You are now at Point K, and the end of the Pattern "A". Return to straight and level, cruising flight. You should now be on the original heading, and at the original altitude.

When you first begin practice in Pattern A it is recommended that the times shown here be doubled. As you gain proficiency, the times can be flown as described. Doubling the times in initial practice gives you more time to think ahead to the next point, and to be ready to accomplish whatever maneuver is required at that point. Once proficiency is established in the Pattern A, you can reduce the times on the legs as described here and shown on the chart—but in no case should the times be reduced further.

Pattern "A" is the last step in instrument flying prior to beginning radio navigation. It gives practice in aircraft control, planning, and timing. If you can do a good Pattern "A", radio navigation on instruments will be no problem.

**Emergency Aircraft Control**

For years the flying fraternity has sponsored a tricky instrument maneuver known as "recovery from unusual attitudes." The general idea behind this delightful maneuver is that you have lost the gyros in your artificial horizon and D.G. (or they tumble, which is the same thing), and you are left with the pitot-static instruments and the old reliable turn and slip indicator whose gyro cannot tumble. You must return your aircraft to straight and level flight via the pitot-static instruments and the turn needle. It requires some adroit flying, and not a little amount of skill.

The common manner in which this maneuver is practised

is for the instructor to fly the airplane while the student closes his eyes. The instructor puts the airplane in some "unusual attitude," then yells, "You've got it!" whereupon the student is supposed to grab hold of the stick and wrestle the craft back to straight and level flight with reference solely to the turn needle and pitot-static instruments. The problem is that the student is suddenly presented with a situation on which his brain has been keeping no time-history. A certain amount of time is lost while the brain observes the instruments, interprets them, and establishes a time-history of their actions. Once the time-history is established, the brain's computer predicts the next move of the aircraft, and if it is not desirable, the brain must compute what action *would* be desirable.

After deciding what action *would* be desirable, the brain computes how to attain this desirable action with the available instruments, i.e. the pitot-static system instruments and the turn needle. When this information is computed, it is transferred to muscle action, and the pilot begins corrective action to right the aircraft. If the pilot's brain is in top-notch condition, and if he has sufficient experience stored in his brain's computer, and if the brain sorts this experience information in the proper manner, and if the instrument interpretation is correct, and if the desirable action is given to the pilot's muscles properly—then the pilot will do the right thing the first time. Unfortunately, there are two many "if's" involved—too many chances for the pilot to end up taking the wrong corrective action.

Practicing recovery from unusual attitudes in the old-fashioned manner is not desirable, nor does it teach anything. First, unusual attitude recovery practice assumes the artificial horizon and D.G. fail, *and* the aircraft immediately assumes some *unusual* attitude that the pilot cannot or does not understand or control. Second, and most important, it assumes no previous time-history.

During instrument flight the brain's computer keeps a running time-history of the flight. It knows what has just happened, what is now happening, and what the pilot desires to have happen next. Any time this time-history is interrupted, the pilot is in danger. This is why it is important to devote only brief glances to charts, maps, and logs during actual instrument flight. Bring such charts, maps, and logs into your normal instrument crosscheck to avoid any major interruptions in the brain's time-history of the flight.

Practice recovery from unusual attitudes assumes a complete breakdown of the brain's time-history function. Even if both the artificial horizon and the D.G. fail in flight, the brain still keeps a time-history! When these two instruments fail, the brain is taxed to keep up with what is going on; but if it has had adequate practice using the turn needle as the sole gyro instrument, there is no reason why the brain should not continue to keep track of things completely enough to maintain desirable aircraft control. For this reason, the old-fashioned custom of having the pilot close his eyes, then instantaneously assume aircraft control when practicing recovery from unusual attitudes is not natural, and may lead to his placing the aircraft in an even *more* "unusual" attitude.

Some instrument practice should be accomplished using only the turn needle, pitot-static instruments, magnetic compass, and engine instruments. The turn needle becomes the "control" instrument for bank, and the altimeter for pitch. Control pressures are applied in the desired direction, and these instruments are observed for the proper reaction. Since the turn needle reacts immediately, bank control is not difficult. Constant attention must be paid to the turn needle, and in rough air, pick the center of the turn needle's oscillations as the position of the needle. Use smooth, easy control pressures, and trim as often as possible.

Pitch control is not as easy, but can be accomplished with practice. The altimeter must be observed closely, and a definite amount of elevator pressure applied and held. Allow the applied pressure to have an effect before changing the pressure. Trim as often as possible, and make smooth control pressure changes. Instead of definite pitch changes on the artificial horizon you have to substitute pressure changes on the elevators. For example, if the altimeter is descending, you must apply back pressure until it stops descending. Then, you must apply additional back pressure to cause it to ascend back to the desired altitude. You must learn to hold a constant amount of pressure for several seconds at a time in order to allow the airplane time to react, and show this reaction on the altimeter.

With the artificial horizon and D.G. inoperative, the pilot should be able to accomplish the following maneuvers:

1. Straight and Level
2. Level Turns
3. Rate Turns to Specific Magnetic Compass Headings
4. A Rate Climb and Rate Descent
5. Spiral Recovery
6. Stall Recovery

Of the above maneuvers, the least important is the Rate Climb and Rate Descent. If the pilot can merely establish and maintain a steady climb or descent at the proper airspeed, and with constant power, he is doing about as well as can be expected. However, a rate descent is highly desirable since the pilot might some day be faced with making a GCA or ILS approach with only a turn needle for a gyro instrument.

In Spiral and Stall Recovery the pilot should not be expected to take over aircraft control in the middle of the maneuver unless he has been observing the instruments up to the time the spiral or stall is well established. The spiral or stall should be established with the student pilot observing the instruments. He should then assume control, and recover to straight and level flight. To recover from a stall, add power, lower the nose, and level the wings, in that order. To recover from a spiral, reduce power, level the wings, and raise the nose in that order. Do not attempt to raise the nose until the wings are level, or approaching level.

# 5. Radio Aids To Navigation

To navigate is to "direct one's course" (Webster) over the surface of the earth. We do this via a method known as "deduced reckoning" (commonly called "dead reckoning") navigation. Dead reckoning navigation is the process of predicting one's course over the surface of the earth by taking into consideration speed of the airplane, direction of the desired course, and the effect of wind. In VFR flight, checkpoints on the earth's surface are used to determine one's position during the process of the flight. Obviously, in IFR flight this procedure is impossible because one cannot see the ground.

Radio aids to navigation have been developed to help the pilot follow his desired course, and to pinpoint his position over the surface of the earth. In addition, some radio aids to navigation are designed to help a pilot make an approach through the "soup" (clouds, rain and other such obstructions to visibility) to a landing. In this chapter we will examine the various radio aids to navigation, and discuss how the pilot uses them.

**The Four-Course Low-Frequency Radio Range**

During the late '20s the four-course low-frequency radio range was developed and refined to the point where it could be used for IFR navigation. The low-frequency range transmits on a 200-to-400 kilocycle frequency, and has four courses on which a pilot can fly to or from the station. All the pilot needs is a low-frequency radio receiver. Due to the simplicity of the required radio receiver, the low-frequency range enjoyed great popularity, and was the basis on which our federal airways were established.

By adjusting the transmitting towers, it is possible to "aim" the four courses of a low-frequency range in any direction.

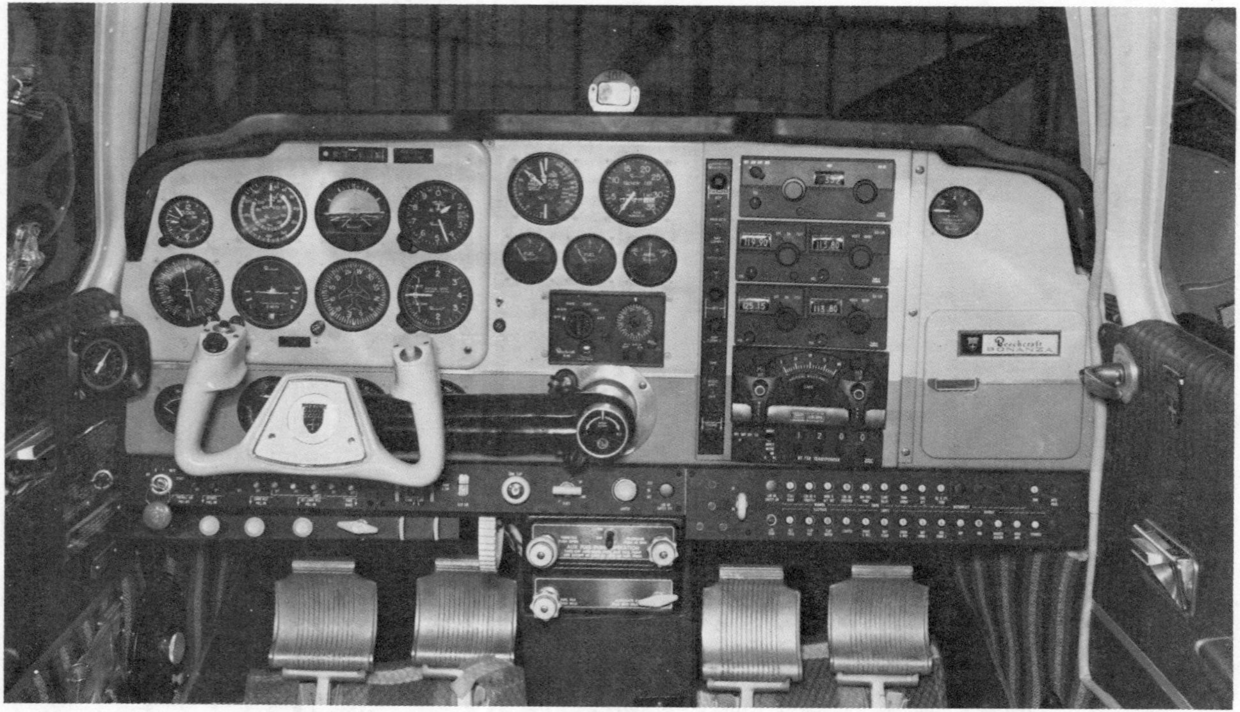

Beechcraft Bonanza instrument panel is orderly and neat. Compare this panel to G17S Staggerwing panel on page 15. (Beechcraft Photo)

Thus, two low-frequency ranges located several miles apart could each have one of their "legs" pointing at each other in such a manner that a pilot could fly outbound on the leg of one range, then switch his receiver over to the other range, and fly inbound to the station on *its* leg (or course). In this manner, the low-frequency ranges delineated the federal airways.

The low-frequency range was the major radio aid to navigation until after World War II at which time the VOR was developed, and began to replace the low frequency range. Today only a handful of low-frequency ranges are left.

One of the major disadvantages of the low-frequency range is the high degree of skill required to "track" on its legs. The range leg (there are four) is determined by listening to the range transmission. When the pilot is flying on a leg, he hears a steady tone. When he drifts off to one side he hears the Morse Code for an A (.–); when he drifts off to the opposite side, he hears the code for an N (–.). When he's on the leg (or "on the beam") the A and N signals overlap to form a steady tone.

There is no way in which the pilot can determine that he is going to or from the station except from the fact that he already knows his approximate position. When his position relative to the station is unknown, he must perform a dandy little maneuver known as the "90° True-Fade Method of Orientation."

The first step in the 90° True-Fade Method of Orientation is to tune and identify the station. The station's identification letters are transmitted approximately every thirty seconds. They are transmitted two times each cycle—first over the N antenna, then over the A antenna. If you are on a range leg, you hear both identification signals with equal intensity. If you are slightly off the leg, you hear one louder than the other. If you're off the leg in an N quadrant, the first set of identification letters is loud, the second (or A set) is weak. The farther you get off the leg, the weaker the A set would become—if you are going into an N quadrant.

When you are well away from the leg in an N quadrant, you hear only one set of identification letters—the first set. Then follows a silent period while the two A antennas transmit the identification letters. In an A quadrant, the silent period would first be followed by the identification letters. As you approach a leg, you begin to hear the opposite quadrant signal identification. This area is called the bi-signal zone because you begin to hear *both* the A and N signals.

The second step in the 90° True-Fade Method of Orientation is to determine the proper "bi-sector heading." If you're in an A quadrant (there are two A quadrants and two N quadrants), you turn the aircraft to either of the A quadrant bi-sector headings (see chart). Obviously, if you're in an N quadrant, you'd turn to either of the bi-sector headings shown in the N quadrant. The bi-sector headings are those headings shown between the range legs with an arrow pointing toward the low frequency range station.

The third step, once you've turned to the bi-sector heading, is to determine if you're flying *to* or *from* the station. This is accomplished by listening to the build or fade in the range signals. The technique is to turn the volume on your receiver down until you can just barely hear any signal at all. Then fly for a minute or so, and try to determine if the signal strength is building or fading. If it's fading, you're flying from the station. Make a turn to the other bi-sector heading for the quadrant (A or N) you're in.

If the signal is increasing in strength, you're flying toward the station—maintain the present bi-sector heading. You will eventually cross a range leg. The fourth step is accomplished when you cross the leg: time the leg. That is, measure the number of seconds it takes you to cross the range leg. If the time is more than seven seconds, proceed to Step Five. If the time is less than seven seconds, turn to the outbound heading of the leg you desire—you're too close to the station to proceed with the 90° True-Fade Method of Orientation—and hold this heading for thirty seconds to one minute, to allow yourself time to fly past the station and definitely be flying alongside the desired leg. Then turn to intercept the leg. This is called a "Close-In Procedure" because it's what you do when you intercept a leg close-in to the station. If the time required to cross the leg is more than seven seconds, proceed to Step Five.

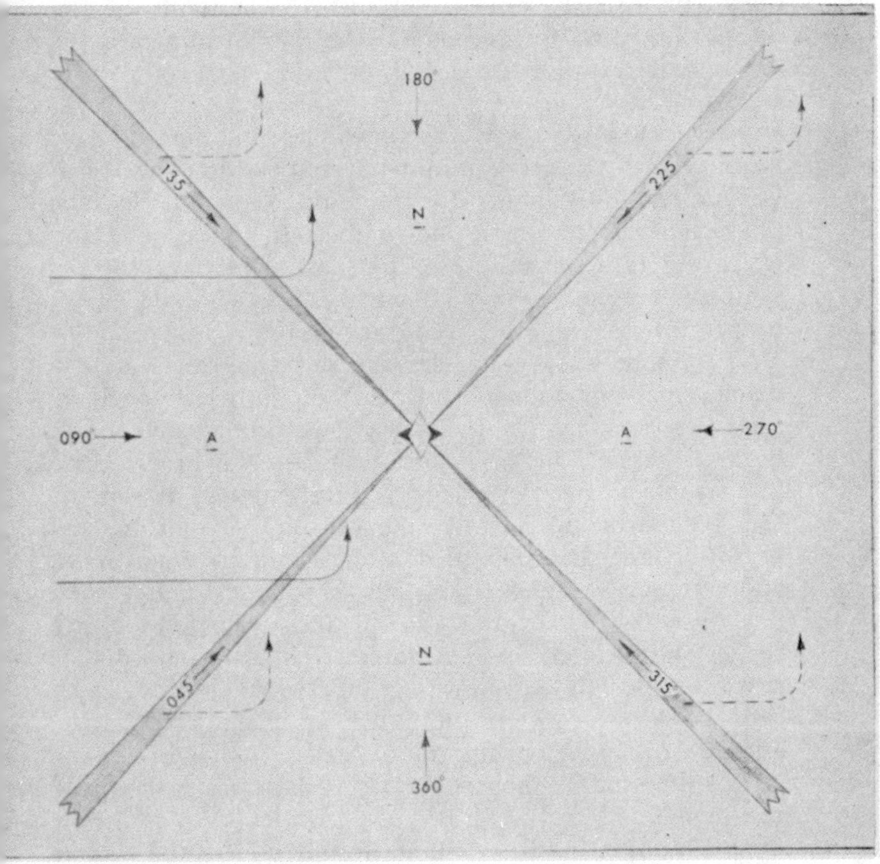


**Four-course low-frequency radio range. Four arrows between range legs pointing toward station are bisector headings of the four quadrants. Solid lines crossing legs illustrate 90° True Fade Orientation; dashed lines illustrate On-Course Orientation.**

Step Five is to cross the leg, and make a 90° left turn. The solid line with an arrow on the accompanying chart shows the track of an airplane in the Western A quadrant. He will intercept either the northwest or southwest leg. When he makes the 90° left turn, he holds the heading for at least one minute. If he intercepted the southwest leg, obviously he'll quickly re-enter the leg. If he intercepted the northwest leg, after flying the new heading for one minute he will not have re-entered the leg, and so realizes it is the northwest leg and makes a 180° turn back towards the leg.

The procedure for any one of the four quadrants is the same. When a leg is entered, begin the timing. If you require more than seven seconds to cross the leg, *immediately* begin a *left* turn when you have crossed the leg, and hold the new heading for at least one minute. If you re-enter the leg, it was the one to the right; if you do not re-enter the leg, the one you crossed was the one to the left of the quadrant. Make a 180° turn and you will shortly intercept the leg again. Once you are on a known leg, it is a simple matter to turn inbound, and begin to track on the leg.

The dashed lines on the drawing show the case of an airplane actually on a range leg when the pilot tunes and identifies the station. Should this happen, turn to any one of the four bi-sector headings. Since the bi-sector headings are at an angle to the range legs, you will quickly fly off the leg. When you do, identify the quadrant (A or N) and begin an immediate left turn. If you're in an A quadrant, and were originally flying a 090° bi-sector heading, you obviously are in the eastern A quadrant. When you leave the leg and turn left, you will intercept it again immediately if you were on the northeast leg; otherwise you'll fly deeper into the A quadrant. In the latter case make a 180° turn, and you will intercept the southeast leg.

The inbound heading of the leg is shown as is the quadrant signal (A or N). Once on a known leg, turn to the inbound heading if you desire to fly inbound to the station. If you desire to fly away from the station, turn to the reciprocal (opposite) of the heading printed on the chart.

Let us assume we desire to track inbound on the southwest leg of the range shown on the accompanying diagram. The heading is 045°. An N quadrant is to our right, and an A quadrant to our left. We fly the 045° heading, and if we drift into an N quadrant, we know we have drifted to the right. So, we turn about 20° to 30° (up to 45° in a stiff wind) back toward the leg. Once we're on the leg again, we adjust our heading 3° to 5° to the left of 045°. This should hold us on the leg. If it doesn't, we turn more to the left of 045° when we get back on the leg. What we are doing is flying a heading into the wind that will make good the

045° track across the ground so that we will remain on the range leg. If the wind were from the right of the leg, obviously we'd drift off into the A quadrant, and have to hold a heading to the right of 045° to remain on the leg.

The width of a range leg is defined largely by the volume of the range signal. The higher you have the volume control, the narrower the range leg. It is desirable to carry the volume as high as possible. If you do, the range leg will be approximately 3° wide.

As you fly toward the station, continue to turn the volume on your receiver down. One of the surest indications that you're approaching the range is a constant adjustment of the receiver volume to keep it down to a low level. As you pass directly over the station, you pass through the "cone of silence." All range signals disappear completely. At that time you are directly over the station. As you fly away from the station, the signals jump rapidly to a high level as you fly out of the cone of silence, then begin to fade rapidly as you continue to fly from the station. As you fly away from the station, you must continually adjust the receiver volume to a higher level to hear the range signals.

**Low-Frequency Homers and Automatic Direction Finding (ADF)**

A low-frequency homer (homing beacon) operates in the same radio frequency range as does the four-course range, but the homer is non-directional, i.e. it transmits a constant signal in all directions. In order to fly to or from a homer, you must have some sort of directional loop antenna attached to your low-frequency receiver. A loop antenna has the characteristic of receiving a loud signal when the plane of the loop is in line with the station, and almost no signal (a null) when the plane of the loop is perpendicular to a line from the loop to the station.

The null position is used for flying to and from the station since the null is narrower than the maximum signal area in terms of degrees of rotation of the loop antenna. The simplest form of loop is manually operated in that you must manually rotate it to find the "null." Using a manual loop you must rotate the loop back and forth over a five-to-ten-degree spread to obtain a usable null. You must also adjust the receiver volume. The higher the volume the smal-

**Components of two Bendix ADF sets. One on right is conventional round dial tuning while control head on left has more advanced digital tuning. Unit in center is the antenna. (Bendix Photo)**

ler the null, which is most desirable. However, as you fly toward the station, you must turn the volume down, or the signals will increase to such a level that the null will disappear.

Since there obviously is a null when you're flying toward the station, and one when you're flying away from it, how do you determine if you're flying to or from a station? You must turn the aircraft so that the null is about 45° off the nose of the aircraft. You listen to the null, and move the loop to keep it on the null position. If the null is moving counterclockwise, the station is in front of you. If the null is moving clockwise, the station is behind you. Turn to put the null on the nose of the aircraft, and fly to the station.

ADF stands for "Automatic Direction Finding," and an ADF set is nothing more than a low-frequency receiver with a standard low-frequency wire antenna, a loop, and a motor to rotate the loop. The ADF set contains a circuit which automatically directs the motor to drive the loop to a null position. It also uses the ordinary wire antenna (called a "sense" antenna) to solve the 180° ambiguity so that the needle on the ADF indicator dial in the cockpit always

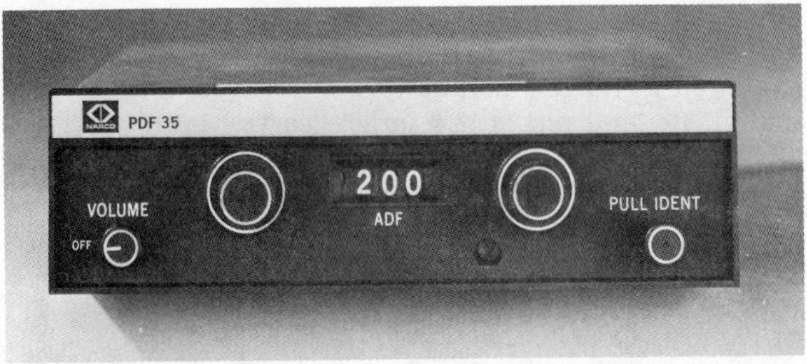


**Narco ADF with digital tuning aids in rapid facility identification. Rotating dial type tuners are more difficult to get on frequency, may cause pilot distraction during a critical flight phase. (Narco Photo)**

points to the station. The needle is arranged so that it will follow the loop's position, and the sense antenna signal is used to tell the needle to point to the station rather than away from it.

Low-frequency homers are used as primary approach facilities for some airports, and as the outer compass locator for an ILS (Instrument Landing System) approach. While the four-course low-frequency range is fast disappearing, the low-frequency homer appears to be in a position to stay with us for a while. While it won't do as much for you as the VOR, it is cheaper to build, and easier to install.

**Very-High Frequency Omnidirectional Range (VOR)**

The old four-course low-frequency range has two major disadvantages. First, it has only four courses to and from the station, and second, static from electrical storms tends to blot out its signal and make it unusable. The VOR was developed to correct these two disadvantages. The VOR offers 360 courses to or from the station and, since it operates in the VHF (Very High Frequency) band, it is relatively static-free and is not affected by electrical storms. The VOR does have one disadvantage that the low-frequency range didn't have, and that is the fact that VHF is "line-of-sight." By line-of-sight we mean the VHF signals travel in a straight line: will not bend to follow the earth's surface, and are not reflected off the Kenny-Heavyside layer to give long range signals. Thus, the VOR is usable only as long as you are "in sight" of it, i.e. no man-made or natural solid intrudes on a straight line drawn between you and the VOR station. Because the earth's surface curves, the further you are away from the VOR station, the higher you must go to receive the VOR signals—due to this line-of-sight characteristic of VHF radio signals.

**Fulton County VOR transmitter. Located on a hill overlooking Fulton County Airport, the VOR transmitter is typical of all VOR stations. Since VOR (like all VHF) is line-of-sight, best locations are on hills or in open country.**

Today the VOR forms the basic radio aid to navigation on the federal airways, and VOR stations are placed close enough together so that the line-of-sight problem is not evident in flying the federal airways. VOR stations transmit from 108.0 to 117.9 megacycles. VHF television transmission begins down around 52 megacycles and extends up to around 90.0 megacycles. FM radio begins above 90.0 megacycles.

The VOR station transmits two signals. One is a fixed reference signal, and the other is a variable signal. The fixed and variable signals are in phase at magnetic north. All VOR stations are oriented to magnetic north. As you proceed clockwise around the station, the variable signal changes position with respect to the fixed signal about as two sine waves change, so that, at a course of 090° from the station, the two sine waves are 90° apart. At a course of 180° from the station, the two signals are 180° apart, and so on around the station until they are again in phase at magnetic north. These signals rotate at approximately 1800 revolutions around the station per minute so that, for all practical purposes, the signal is constant at any one point around the station.

The VOR signal varies around the station so that there are 360 different signals around the station, equally spaced. These signals form "radials" which are *magnetic courses outbound from the station.* The 090° radial goes out from the station toward the east, the 180° radial goes out toward the south, etc. all around the station. The VOR receiver in your aircraft measures the fixed (reference) and variable signal, and determines which radial you are on.

You are provided with a course selector by means of which you can select any desired course to or from the VOR station. A Course Deviation Indicator needle is mounted on the VOR set to tell you whether you should turn right or left to intercept the course you have set. When the CDI (Course Deviation Indicator) needle is centered, you are on the desired course or radial. A course and a radial are not the same thing. A radial is a magnetic course outbound from the VOR station. A course is merely a desired magnetic course over the ground. The VOR receiver has an ambiguity meter which reads To or From. If the course you select on the course selector (or omni-bearing selector—the terminology depending on the brand of set you have) will take you to the station, the ambiguity meter will indicate To. If the course selected will take you from the station, the ambiguity meter shows From.

When you select the desired course, you must turn the aircraft toward the CDI (to the right if the CDI is to the right of center, to the left if the CDI is to the left of center) in order to fly toward the selected course. When you arrive on the selected course, the CDI will center. You must now adjust the aircraft's heading to fly the course. The heading of the aircraft will be the same as the selected course if there is no wind. Otherwise, you may have to head the aircraft into the wind to keep the CDI centered.

When you cross the VOR station, the most positive indication that you are crossing the station is the ambiguity meter going from To to From. Once you pass the VOR station, the meter will show From, and the CDI will continue to show whether you are right or left of the selected course.

If you desire to turn around, and fly back to the station, you should turn the course selector around to the new course. Otherwise, the CDI will indicate backwards. That is, it will move off the left side of center if you are to turn right to get back on course! For this reason, you should always keep the course selector within 90° left or right of the course you desire to fly, so that the CDI will indicate properly.

**Tactical Air Navigation (TACAN)**

Very much akin to VOR is the military Tactical Air Navigation system commonly known as TACAN. TACAN operates at around 1,000 megacycles, and there are no known civil radios for the TACAN system. Since the course signals for TACAN give the pilot identical indications to those we have just described for VOR it is unlikely that the more expensive TACAN receivers will ever become popular with civil users. TACAN offers no superior advantage—VOR does the same job and with a less expensive receiver. One thing TACAN does have is DME (Distance Measuring Equipment) which is usable on civil DME radios.

The DME radio transmits a signal which the ground station picks up and duplicates. It returns the signal to the aircraft receiver which changes it to read in nautical miles, thus giving the aircraft's distance to the DME station. On civil DME equipment, you tune the associated VOR frequency to receive the proper DME signal. VOR frequencies are paired with TACAN frequencies for use on civil DME radios. Some civil DME equipment will show not only the distance to the station, but also the aircraft's ground speed! Naturally, for the ground speed to be accurate, you must be flying directly to or from the DME station.

In the Federal airway system, TACAN stations are being mounted on the same sites as VOR stations. The reulting facility has a complete operation of the civil VOR plus the military TACAN, and military/civil DME. For the military pilot to obtain DME, he must tune in the proper TACAN frequency. The civil pilot using civil DME equipment, merely tunes in the VOR frequency on his DME equipment for the VOR station located at the TACAN site. Such common locations of VOR and TACAN are

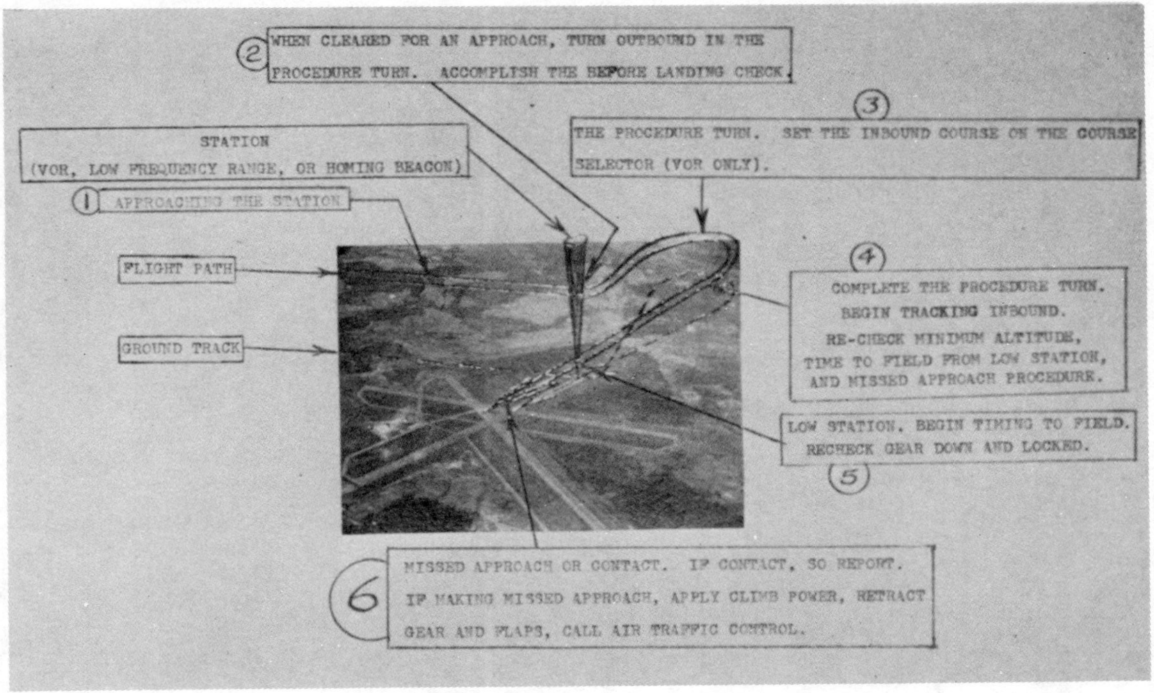


**Standard instrument approach shows flight path arriving at station, entering procedure turn, and making approach. The six steps listed are applicable to most approaches using VOR, TACAN, Low Frequency Range, or ADF (Homing Beacon). Individual situations will dictate changes in procedure occasionally.**

known as VORTACs. Thus, a VORTAC station is a full VOR and a full TACAN station, plus civil DME, all located at the same spot on the ground.

**Marker Beacons**

The marker beacon is a short-range, highly directional beacon which transmits on 75 megacycles. Its beam is aimed upwards, and its signal is dashes or dots, or a combination thereof. Its signal can be heard (audio) or seen (lights) depending on the type of marker beacon receiver you have. Marker beacons are used to establish the point to begin an ILS approach, and are used as the inbound, minimum altitude fix on the ILS system. When used thus with the ILS system they are known as outer and middle markers.

Outer markers on the ILS system transmit a series of dashes at 400 cycles per second. Middle markers transmit alternate dashes and dots at 1300 cycles per second. Marker beacons are so weak in power and so directional in nature, that you must fly almost directly over one in order to receive its signal.

**Instrument Landing System (ILS)**

Using only a VOR or low-frequency radio range, a pilot is required to have a good ceiling and visibility on the airport before he can make an instrument approach. While a takeoff can be made (by a very sharp or very brave pilot, depending on your viewpoint) when the ceiling and visibility are completely zero, it takes a mighty fine system, and a very sharp pilot (whether he's brave or not) to make an approach when the ceiling and visibility are low.

When using the low-frequency range a ceiling of 300 to 400 feet is needed for a safe approach. In some locations the required ceiling will be higher due to local conditions. A VOR approach generally requires almost the same ceiling unless the VOR range is near the airport in which case the limits can be down to around 300 feet of ceiling and one mile visibility.

In order to enable airplanes to land when the ceiling is down around 200 feet and visibility is one-half mile or so, the ILS (Instrument Landing System) was developed. Its history reaches back to just prior to World War II which makes it about 10 years older than VOR. The ILS transmits one course which is of a highly refined nature. This course (called the localizer) operates from 108.1 to 111.9 megacycles on odd tenth decimal frequencies. Even tenth decimal frequencies are VOR in this range. In the 108.1 to 111.9 range there are 20 ILS localizer frequencies.

Like the low-frequency range, homing beacons, VOR, and TACAN (not DME), the ILS system transmits a coded identification. It consists of three letters of the Morse Code in front of which is transmitted the letter "I" to indicate an ILS system. This identification is transmitted only on the localizer frequency. Paired with the localizer frequency are 20 UHF (Ultra High Frequency) glide slope frequencies. They are:

| Localizer | Glide Slope | Sequence |
|---|---|---|
| 110.3 | 335.0 | 1 |
| 109.9 | 333.8 | 2 |
| 109.5 | 332.6 | 3 |
| 110.1 | 334.4 | 4 |
| 109.7 | 333.2 | 5 |
| 109.3 | 332.0 | 6 |
| 109.1 | 331.4 | 7 |
| 110.9 | 330.8 | 8 |
| 110.7 | 330.2 | 9 |
| 110.5 | 329.6 | 10 |
| 108.1 | 334.7 | 11 |
| 108.3 | 334.1 | 12 |
| 108.5 | 329.9 | 13 |
| 108.7 | 330.5 | 14 |
| 108.9 | 329.3 | 15 |
| 111.1 | 331.7 | 16 |
| 111.3 | 332.3 | 17 |
| 111.5 | 332.9 | 18 |
| 111.7 | 333.5 | 19 |
| 111.9 | 331.1 | 20 |

The ILS localizer defines the centerline by the overlapping of a 90-cycle and 150-cycle tone. Flying inbound on the ILS approach course, the 150-cycle tone is to the right and is called the "Blue" sector. The left sector is the "Yellow" sector, and has the 90-cycle tone. If your VOR receiver has localizer function, then its CDI will operate

**Compact Narco navigation receiver for VOR and ILS has tuning function right on face of navigation indicator. (Narco Photo)**

on ILS (when you select LOC function) just as it does on VOR, but with much greater sensitivity. Also, when you turn around and go back on the ILS, you cannot turn the course selector around (it has no effect on ILS) to make the CDI work properly. You simply must remember that you turn right when the CDI is left when you're flying backwards on the ILS.

The ILS glide slope operates a horizontal GSI (Glide Slope Indicator) needle in the same manner that the localizer operates the CDI. By keeping the GSI centered, you will fly down the glide slope to the runway. If the GSI needle is above center, you must pull up (decrease rate of descent) to get back on the glide slope. If the GSI needle is below center, you must increase your rate of descent.

There will normally be an OM (Outer Marker) at the point where you intercept the glide slope and an MM (Middle Marker) at the contact flight point on the glide slope. At either or both markers there may be an ADF homing beacon. If so, the approach plate will have an "L" in front of OM or MM, and the homer frequency shown in a box nearby with its identification.

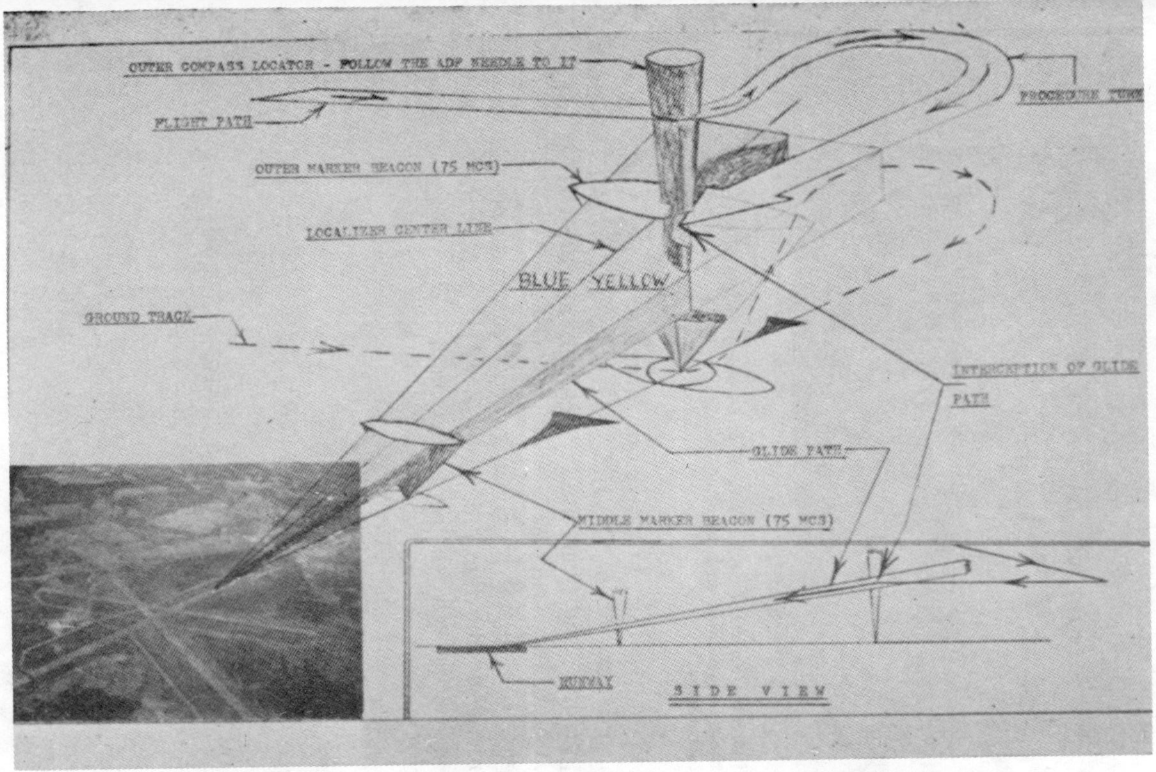


**Instrument Landing System showing marker beacons and outer compass locator. Flight path for a typical approach is shown with a small side view illustrating point (outer marker beacon) where glide path is normally intercepted.**

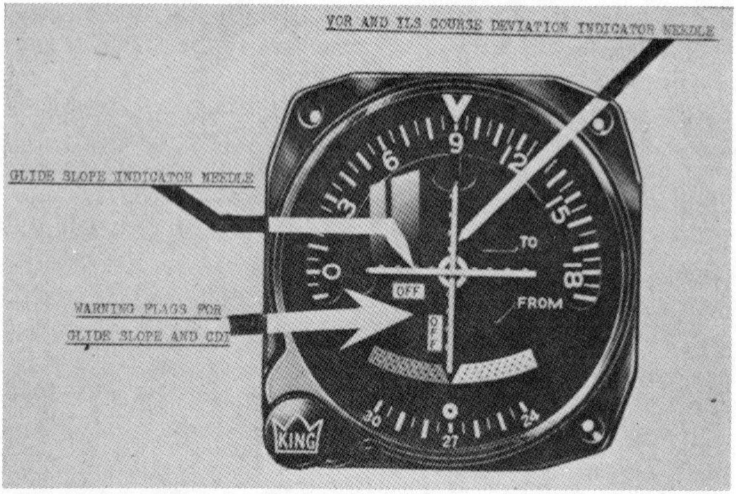


**Indicator for VOR and ILS with glide slope needle. Course selector is the azimuth dial with selected course displayed under "V" at top of instrument.**

It is possible to make an instrument approach while going backwards on the ILS. This is known as "Back Course Approach," and you must remember that the CDI is backwards. When it swings left, you turn right. When it swings right, you turn left. Since no back-course approach has a glide slope the minimum ceiling and visibility for a back-course approach will be higher than for a normal ILS approach.

## 6. Publications to Aid the Pilot

The federal government has several publications which are essential for the instrument pilot. First, he should subscribe to the *Airman's Information Manual* Part 3 and 3A — Operational Data and Notices to Airmen as an absolute minimum. He should also have Part 1 — Basic Flight Manual and ATC Procedures, Part 2 — Airport Directory, and Part 4 — Graphic Notices and Supplemental Data. The prices are $4.00 for an annual subscription to Part 1, $4.00 for Part 2, $20.00 for Part 3 and 3A, and $1.50 for Part 4. They are available from:

Superintendent of Documents
Government Printing Office
Washington, D.C. 20402

Enroute, Terminal Area, and Instrument Approach Procedure Charts are published by National Ocean Survey, Distribution Division, Riverdale, Maryland 20840 and prices vary. Write for their list of charts available.

**Enroute Charts**

At the present time the airways in this country are divided into two layers:

1. Low Altitude—surface to, but not including, 18,000 feet.
2. High Altitude—at and above 18,000 feet.

A set of navigation charts primarily for IFR flights is printed for each stratum (layer) of the airways. They are called the *Enroute Charts.*

Using a scale of miles similar to a *World Aeronautical Chart,* some 30 *Low-Altitude Enroute Charts* cover the entire United States (except Alaska and Hawaii). The charts are printed back-to-back with consecutive charts on the same piece of paper. There is a generous overlap between charts. The charts have a section on one end devoted to a legend, a map of the United States showing the individual chart in bold outline, frequency data, arrival/departure chart data, and the effective dates of the chart. Since aeronautical information (for some reason) changes constantly, the charts are reprinted approximately every four weeks to keep current with the latest changes.

The *Enroute Charts* contain all airway information, and information on radio aids to navigation that define the airways. VOR and VORTAC frequencies and call signs are shown, as are low-frequency range stations, and most ILS systems. The magnetic course of all airways is shown, and

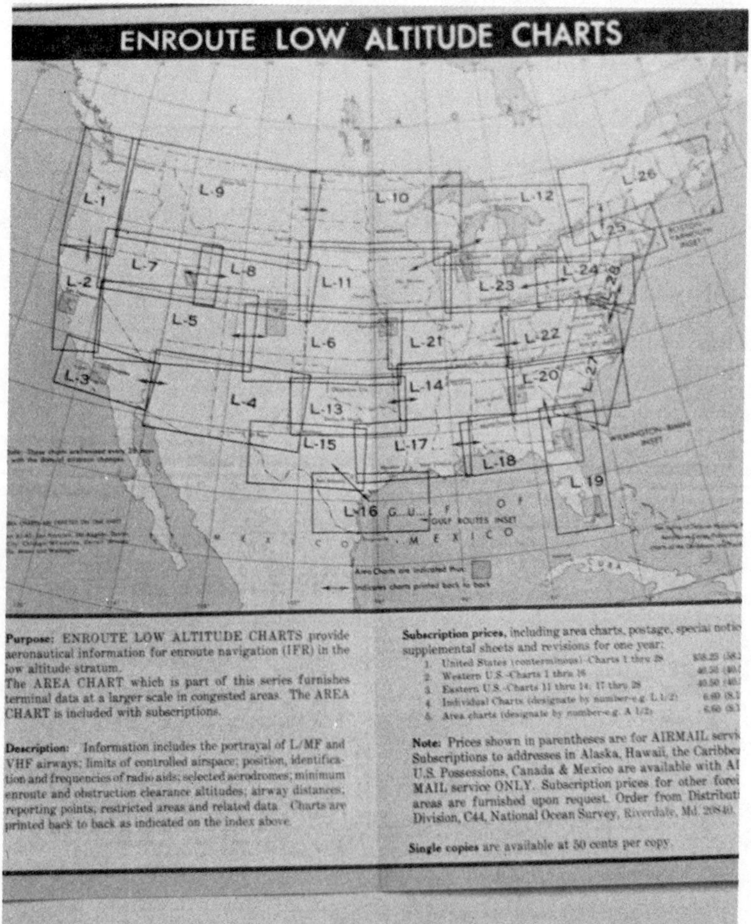


**Map of United States showing Enrout Low Altitude chart coverage. These are published by government, others are available from private printers.**

distances between facilities and reporting points is shown in nautical miles.

**Arrival/Departure Charts**

Published for selected terminal areas in the United States, the *Arrival/Departure Charts* expand the information for those areas. They are more detailed than *Enroute Charts,* and are much easier to use in a terminal area. *Arrival/Departure Charts* are published for approximately 33 major terminal areas in the United States.

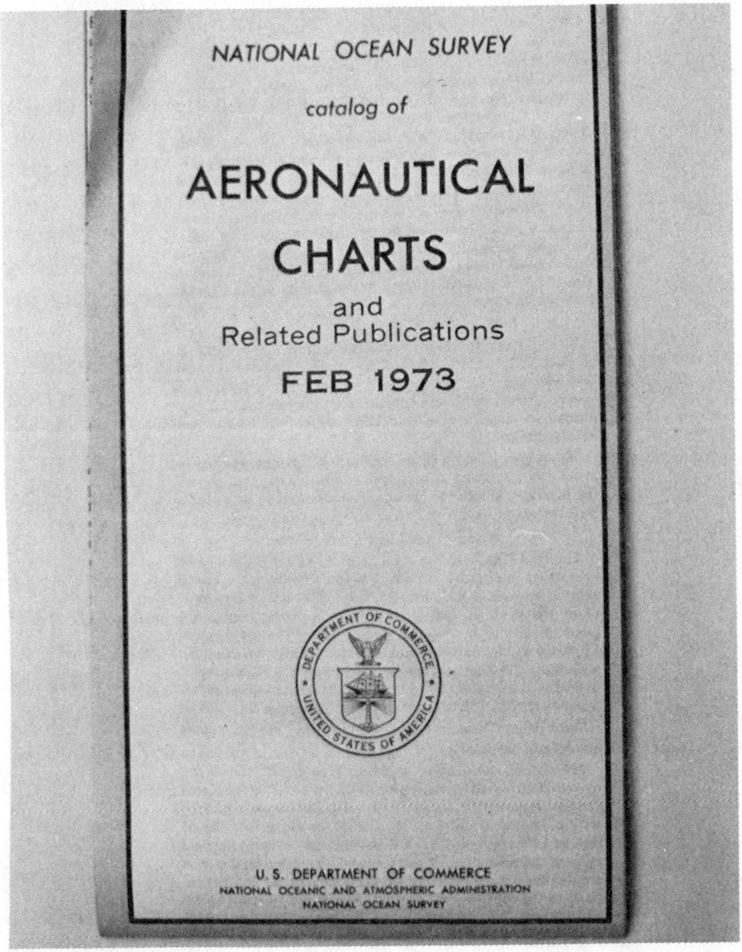


**National Ocean Survey catalog lists all available aeronautical charts and prices. Write them at "Distribution Division (C-44), National Ocean Survey, Washington, D.C. 20235" and ask for their catalog of aeronautical charts.**

**Instrument Approach Charts**

At the end of an IFR flight, a pilot must descend and make an approach to the airport where he desires to land. The procedure for making such an approach is published on what is known as the *Instrument Approach Charts.* A plan view of the approach procedure is given to show location of necessary navigation aids, headings and courses to fly, and altitudes for emergency use. A profile view of the procedure shows proper and normal altitudes for the approach. Altitudes shown on the profile are *minimums* for each point in the procedure. To descend below them is unsafe, and foolish.

An airport diagram is shown, and the Jeppesen charts have a rather detailed airport diagram which is printed on the reverse side of the approach procedure. All necessary radio frequencies are shown on the *Instrument Approach Charts,* and a table gives the various minimum ceilings and visibilities for the approach procedure. A pilot should carefully study the *Instrument Approach Chart* before attempting an actual instrument approach. One should never try to memorize the information shown on these charts.

Normally the *Instrument Approach Charts* are carried in loose-leaf, ring binders. The C&GS offers only one style binder, but Jeppesen sells plain plastic ones, and nice, leather binders on which your name can be imprinted. The charts are revised whenever necessary to reflect current, up-to-date information.

**Airman's Guide and Flight Information Manual**

Both the *Airman's Guide* and *Flight Information Manual* are federal government publications. Information pertaining to both VFR and IFR flight on Air Traffic Control procedures, standard broadcast radio stations (for ADF use), VOR checkpoints, and other general helpful facts are contained in the *Flight Information Manual.*

Perhaps the most valuable manual available to any pilot, VFR *or* IFR, is the *Airman's Guide* published by the U. S. Government Printing Office. Issued in two sections, the main portion is printed every two weeks and the Directory of Airports and Seaplane Bases as necessary. The main section of the *Airman's Guide* contains Notices to Airmen (NOT-

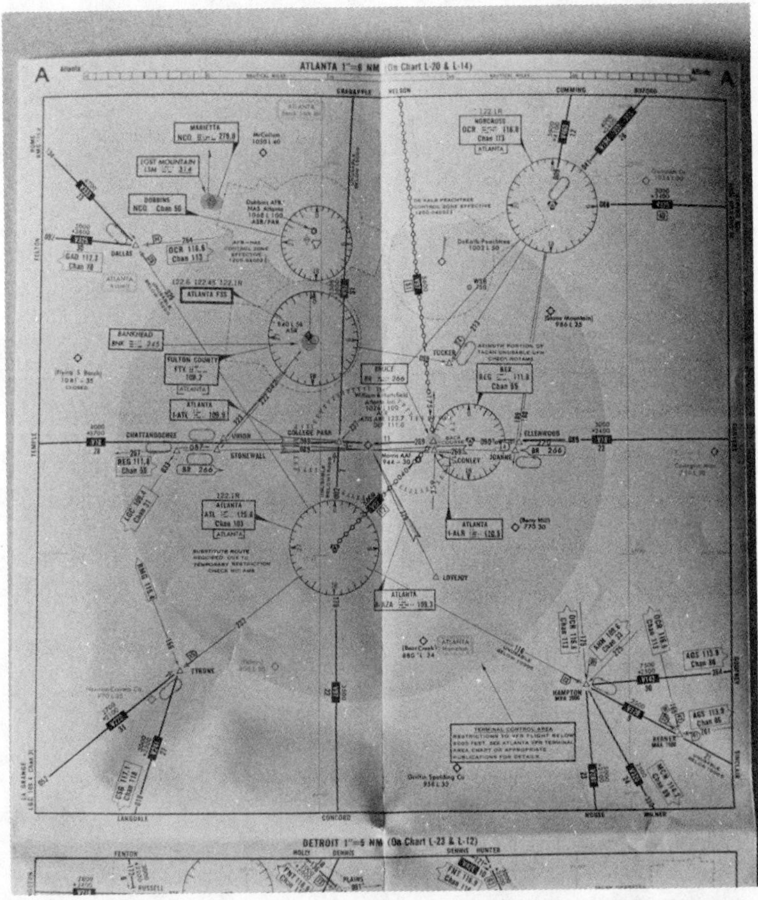


**The Atlanta, Georgia area arrival chart. These charts contain a blow-up of congested terminal areas and show many facilities not depicted on the enroute charts so their procurement and use is a must for instrument pilots. There are presently 10 area charts, all depicted on one sheet.**

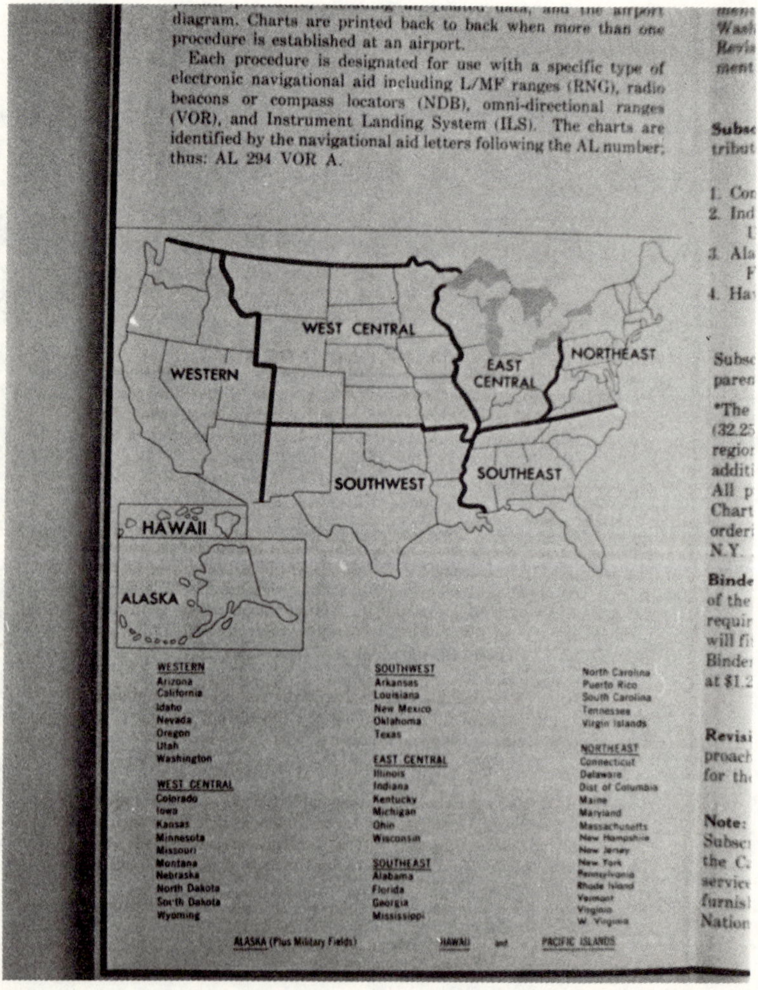
diagram. Charts are printed back to back when more than one procedure is established at an airport.

Each procedure is designated for use with a specific type of electronic navigational aid including L/MF ranges (RNG), radio beacons or compass locators (NDB), omni-directional ranges (VOR), and Instrument Landing System (ILS). The charts are identified by the navigational aid letters following the AL number; thus: AL 294 VOR A.



WESTERN
Arizona
California
Idaho
Nevada
Oregon
Utah
Washington

WEST CENTRAL
Colorado
Iowa
Kansas
Minnesota
Missouri
Montana
Nebraska
North Dakota
South Dakota
Wyoming

SOUTHWEST
Arkansas
Louisiana
New Mexico
Oklahoma
Texas

EAST CENTRAL
Illinois
Indiana
Kentucky
Michigan
Ohio
Wisconsin

SOUTHEAST
Alabama
Florida
Georgia
Mississippi
North Carolina
Puerto Rico
South Carolina
Tennessee
Virgin Islands

NORTHEAST
Connecticut
Delaware
Dist of Columbia
Maine
Maryland
Massachusetts
New Hampshire
New Jersey
New York
Pennsylvania
Rhode Island
Vermont
Virginia
W. Virginia

ALASKA (Plus Military Fields) HAWAII and PACIFIC ISLANDS

**Instrument approach procedure charts published by the National Ocean Survey are broken down into areas of the United States and each area is contained in a separate flip chart book.**

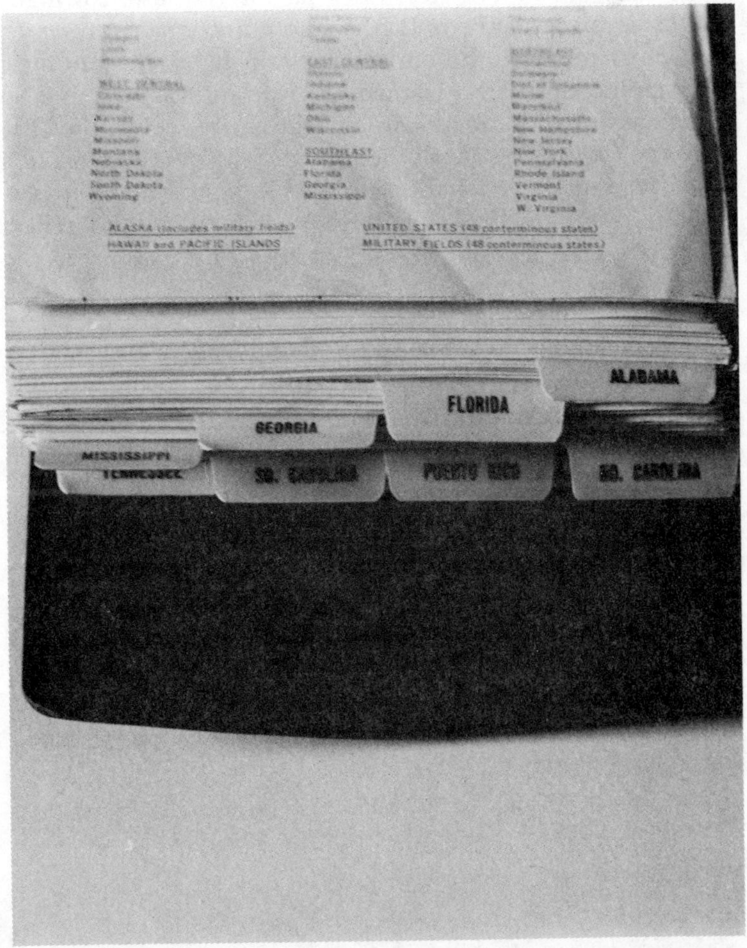


**An instrument approach procedure chart book showing tabs with charts arranged by state. A subscription covers revisions for one year, but individual airport charts may be purchased separately.**

AMS), and a Radar-Radio facility listing section. We recommend that an up-to-date *Airman's Guide* be carried on every VFR and IFR flight.

**Sectional Aeronautical Charts**

Drawn to a scale of 1 to 500,000 (approximately 8 statute miles per inch), the *Sectional Aeronautical Charts* cover the Zone of the Interior and Hawaii. Primarily for use in VFR flight, they are nevertheless essential for IFR use. No flight should be made, VFR or IFR, without *Sectional Aeronautical Charts* to cover the entire route of flight. Very minute topographic details are shown. Airway information is also shown, as are all radio aids to navigation. However, the IFR *Enroute Charts* previously described (either C&GS or Jeppesen) should be used for actual IFR flight since the sectional charts are not revised frequently enough to reflect up-to-date information on radio aids to navigation or the federal airway system.

**Industry Publications**

Many manuals and books are offered by private industry to aid a pilot in obtaining his instrument rating, or any other rating he might desire. One of the most outstanding manuals for the instrument pilot is the Weems and Zweng manual titled *Instrument Flying.* It is available from the Pan American Navigation Service in North Hollywood, California. Jeppesen and Company also publishes a home study guide manual for the instrument rating. The manual is available directly from Jeppesen or through one of their authorized dealers—Pan American Navigation Service is a major one.

## 7. Air Traffic Control Procedures

Instrument flight on the Federal airways or in Federally controlled airspace is under supervision of an arm of the FAA (Federal Aviation Agency) known as the Air Traffic Control Division. In the field they are known as "ATC." These birds have the United States divided up into thirty areas or "ATC Centers." Each Center (properly an "Air Route Traffic Control Center," or ARTCC) is located in one large building from which is controlled IFR traffic in the area. The Center ("Center" is standard terminology for the Air Route Traffic Control Center) controls IFR traffic within its area which is actually en route. Traffic departing an airport is handled by Departure Control; traffic arriving is handled by Approach Control, and traffic transitioning from an instrument approach to contact flight for a landing is handled by the airport tower.

All of these agencies of the FAA work in close coordination to insure a smooth flow of IFR traffic from takeoff until touchdown at destination. Traffic flow is *not* always smooth, but the fact that it is smooth most of the time is something closely akin to a miracle. We must give the individual Air Traffic Controllers credit for having performed this miracle. Pilots who know ATC procedures help, but the burden of the job is on the ATC controller.

The *Flight Information Manual* published by the Government Printing Office is helpful on ATC procedures, and instrument manuals offered by private industry deal extensively with the subject. We personally recommend a book published by Sports Car Press entitled *Guide to Air Traffic Control.* While this work was intended primarily for the VFR pilot, it has such a detailed breakdown of the FAA air traffic control system that it is extremely valuable to the IFR pilot also.

In instrument flying, your contact with ATC will be primarily via the radio. However, we strongly recommend that you personally visit your local control tower and Center, if you can. You will find the controllers friendly, helpful, and extremely well informed on what a pilot does. Many of them are pilots, and all of them are interested in helping you learn more about their side of the business.

ATC's primary job is to provide traffic control over IFR traffic in order to insure safe separation between individual airplanes. A secondary job is to give inflight assistance to pilots and assistance in emergencies.

**Filing a Flight Plan**

Your first contact with ATC will be a "silent" contact when you file your flight plan with an FSS (Flight Service Station). You can file the flight plan in person or by telephone. It is permissible to file an IFR flight plan in the air, but it is not considered very professional to do so. (It is also not considered very professional to file a VFR flight plan in the air if you could have done so in person or by telephone before takeoff). When the FSS receives your flight plan, they give it to the center by telephone.

The Center receives your flight plan, makes up a "flight strip" on you, and puts you in the list of airplanes desiring to take off and fly IFR. When it is your turn, or when they can let you enter the inflight stream of IFR traffic, they let you take off. This permission is given you through the local control tower, which also gives you your flight clearance.

**The IFR Flight Clearance**

Before you can take off and fly IFR, the ATC people want to know that you fully understand your "clearance." In your flight plan you specified the route of flight you desire, the altitude you desire to fly, etc. But, ATC may have to alter your route slightly, or give you a different altitude to fly. Also, they may have detailed departure instructions to give you to help you take off and avoid other traffic.

In order to be sure that you understand your clearance, ATC has the tower read it to you over the radio before you take off, and you must read it back, *verbatim,* to the tower.

This insures that you know what you are cleared to do. The clearance will almost always be read in the following sequence:

1. Your destination (which you already know), your major route of flight, and the altitude you are to maintain on the route of flight.

2. Climb instructions, if any.

3. Frequencies on which to contact the departure control agency, and the time to contact them.

A sample clearance might run like this:

> ATC CLEARS BEECH FOUR FOUR GOLF TO THE MACON AIRPORT VIA REX VOR, VICTOR FIVE EAST MCDONOUGH VORTAC, FLIGHT PLAN ROUTE. MAINTAIN FIVE THOUSAND. RIGHT TURN AFTER TAKEOFF, CONTACT ATLANTA DEPARTURE CONTROL ON ONE, ONE, NINE, DECIMAL EIGHT AFTER TAKEOFF. MAINTAIN THREE THOUSAND UNTIL VICTOR FIFTY-ONE EAST. CONTACT FULTON COUNTY TOWER ON ONE, ONE, EIGHT, DECIMAL FIVE WHEN READY FOR TAKEOFF.

Now, let's take this clearance apart and see what it means. First, "ATC CLEARS BEECH FOUR FOUR GOLF TO THE MACON AIRPORT . . . " This phrase identifies you, and establishes your clearance limit fix—you are cleared to fly to the radio facility serving the Macon Airport. You are not yet cleared to make an approach to the Macon Airport, but since you're not off the ground yet, it really doesn't matter! When you arrive at the facility serving Macon Airport, you will have clearance to make an approach. If you do not, you are expected to hold on the facility, and await further instructions.

To "hold" you arrive over the facility (radio aid to navigation) and begin a right turn. You make the turn 180°, then fly for one minute. After one minute, you turn back to fly inbound to the facility. Once established on an inbound track to the facility, begin to time the leg inbound. It should be one minute. If it isn't, then vary the outbound leg on the next go-'round to make the inbound leg one minute in length. You are now in the "race-track" type holding pattern awaiting further clearance.

". . . VIA REX VOR, VICTOR FIVE EAST MCDONOUGH VORTAC, FLIGHT PLAN ROUTE." This establishes your route as the one you filed. ATC is going to clear you on the exact route you filed, and they list enough fixes (Rex VOR, and McDonough VORTAC) to clearly establish the route of flight. The phrase "FLIGHT PLAN ROUTE" means you are cleared to fly the route you filed from McDonough VORTAC to the facility serving Macon Airport. It does *not* include approval of the altitude you filed for.

"MAINTAIN FIVE THOUSAND." You are to fly at 5,000 feet along your route of flight.

"RIGHT TURN AFTER TAKEOFF . . ." As soon as possible after you take off, you are to establish a right turn. It is assumed some time will be used up in climbing to 300 to 500 feet before you begin the turn; and in getting your landing gear retracted, setting climb power on the engine, and other such items one ought to do just after takeoff. Never sacrifice aircraft control in order to talk to ATC! Or, to comply with their instructions! Control the airplane *first,* or you might not be around to comply with ATC instructions!

". . . CONTACT ATLANTA DEPARTURE CONTROL ON ONE, ONE, NINE, DECIMAL EIGHT AFTER TAKEOFF." The most important thing here is to be sure you have a definite climb established, and the aircraft under control *before* you attempt to switch the voice radio to 119.8 megacycles to talk to Atlanta Departure Control. When you do, Atlanta Departure will give you such additional instructions as the immediate situation dictates. If they have you on radar and vector you off to one side of your cleared route of flight, follow their instructions. Should you then lose radio contact, return to the cleared route of flight by the most direct method.

"MAINTAIN THREE THOUSAND UNTIL VICTOR FIFTY-ONE EAST." On your route of flight you have been assigned five thousand feet as an altitude to maintain, but this phrase indicates you are to level off at 3,000 after takeoff, and hold 3,000 until you arrive at airway "victor fifty-one east." It further means that, once you arrive at airway victor fifty-one east, you are cleared to climb from 3,000 to 5,000. When you depart 3,000, you are required to call on the radio, and

**Beech Four Four Golf employed in ATC clearance in text is a Beechcraft G17S owned by D. W. Yarbrough of Atlanta, Ga.**

tell ATC you are leaving 3,000. You should say: "BEECH FOUR, FOUR GOLF, VICTOR FIFTY-ONE EAST, OUT OF THREE FOR FIVE." ". . . OUT OF THREE FOR FIVE." means you're leaving 3,000 feet for 5,000. You do not have to report reaching 5,000.

"CONTACT FULTON COUNTY TOWER ON ONE, ONE, EIGHT, DECIMAL FIVE WHEN READY FOR TAKEOFF." This is a tower instruction. Modern airport towers have a ground control frequency and a tower frequency. (See *Guide to Air Traffic Control* in the Modern Aircraft Series). Clearances are given to airplanes about to depart on the ground control frequency. At very busy airports, there is a clearance delivery frequency set aside for the delivery of ATC clearances to departing aircraft. Normally, the tower will instruct you when to contact the clearance delivery to obtain your ATC clearance. The tower instruction mentioned in our sample means you are to switch your radio to 118.5 megacycles when you've finished your engine run-up, and are ready for an immediate takeoff.

**Enroute Procedures**

Once airborne, you contact the proper departure control agency (if one is available), and they will give you instructions. Normally they are concerned only with getting you established on an airway at which time they turn control of you over to the center. When you are instructed to contact the center, you may consider yourself *enroute!*

Once enroute you concentrate on holding the assigned altitude, and flying along the proper airway. Monitor the voice by keeping the volume high enough to hear the center should they call. As you fly along the airways you are required to make position reports at specific points. Those points on the federal airways where you must make a position report are shown as "compulsory reporting points" on the *Enroute Charts* and *Area Arrival/Departure Charts.* Other reporting points are "on request" or "non-compulsory" unless ATC asks you to report over one of them at which time they become compulsory for you. Also, any point you have listed in your flight plan is compulsory for you.

When you arrive at a compulsory reporting point, you must make a position report using such typical identifying phrases as the following:

Aircraft Identification ................ "BEECH FOUR FOUR GOLF"
Position ........................................ "REX OMNI . . . "
Time ............................................. "ONE FIVE . . . "
Altitude ........................................ "FIVE THOUSAND . . . "
ETA for next compulsory
reporting point ........................... "MCDONOUGH AT TWO ZERO . . . "
Next compulsory reporting
point after that ........................... "ATHENS . . . "
Remarks, notes on weather, etc..... "IN CLOUDS, LIGHT RAIN"

Note that time is given in minutes after the hour. For example, 10:15 would be given as "one five." It is neither necessary nor desired to say "ten fifteen," or "one zero one five." Just give the minutes after the hour, and you don't need to add "past the hour"—the ATC controller knows full well the time is past the hour. Altitude is given simply as the altitude without the word "feet" following the altitude. The ATC Controller knows that we measure altitude in feet in this country.

An important item about position reports is to speak normally, neither too fast nor too slow. Remember that the ATC Controller is a human being who must write down what you say. Don't say too much. The example position report takes approximately nine seconds. If you add extraneous words such as "feet," "past the hour," etc., the time required to give the report increases. If your reports take a lot of time, other pilots must wait. Likewise, if other pilots talk too much, you must wait. Sauce for the goose is sauce for the gander, so be brief and give the other fellow a chance. Monitor the frequency before you transmit. Do not talk when someone else is talking. You'll block out his transmission, and neither of you will be heard. You can't out-shout somebody on a radio frequency!

For radio communication technique we highly recommend Chapter Ten in the book *Guide to Air Traffic Control*. Tuning radios, normal voice communications, mike technique, etc. are discussed in some detail, and there is a sample VFR flight voice call. Also available are several phonograph records of actual voice communications—invaluable to the new instrument pilot.

While you are en route, your major voice communications are with the ATC Center. Should you desire information such as weather, or NOTAM information while in flight, do not ask the Center. Instead, advise them you'll be off their frequency for a minute, switch your radio to the nearest Flight Service Station and get the information from them. For weather information, you should listen to the scheduled weather broadcasts on the VOR. These are at :15 and :45 after each hour, so for crying out loud, don't call a Flight Service Station at those times and ask for the weather! The fellow giving the weather is probably the man who will answer you, and he just gave the weather over the VOR frequency. All you had to do was listen to it! However, at other times, or for specific information, don't hesitate to call the FSS (Flight Service Station). But, don't burden the Center controller with weather or NOTAM (Notices to Airmen) requests. The Center controller's primary job is controlling IFR traffic, and if traffic is heavy, he simply doesn't have time to help you. That's why you should call the Flight Service Station.

Should it become necessary to file a flight plan while in the air (either IFR or VFR), do *not* contact the Center. Call the nearest FSS. Once you file the flight plan with the FSS, they will give it to the Center via telephone, and will get your clearance, or have you contact the Center for clearance. Just follow whatever instructions they give you.

**Arrival**

As you approach your clearance limit fix, ATC will probably give you necessary arrival (approach) instructions. If they do not, hold on the inbound course to the fix. At airports where there is an Approach Control facility, you will be told to contact them. Otherwise, you will probably contact the tower, or local FSS if there is no tower. They will tell you whether to hold, or to begin your approach, and should give you the approach procedure you are to use—VOR, low-frequency range, ILS, etc. If GCA (Ground Con-

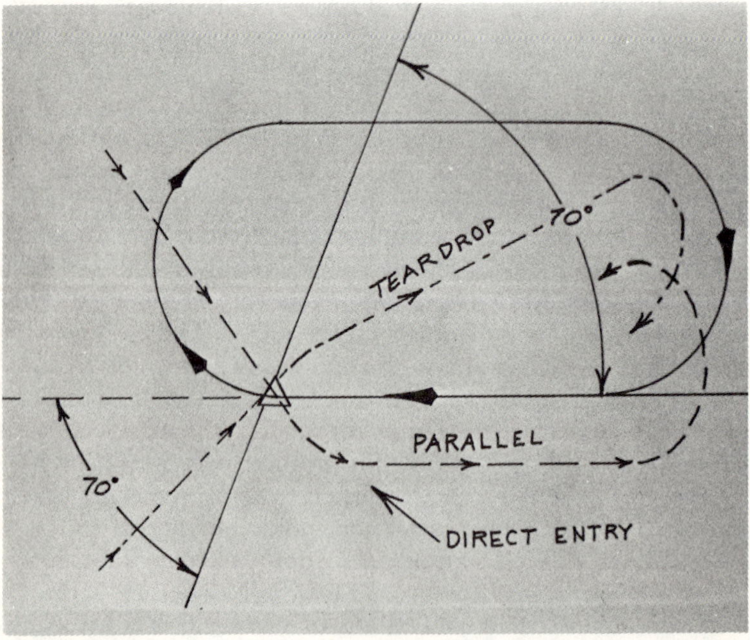


**Recommended holding pattern entries. Solid line with arrows depicts holding path. Dashed lines show teardrop and parallel entry methods, and notation "direct entry" means any entry which is aligned from 110 degrees to the non-holding side around to 70 degrees on the holding side of the inbound holding course may be a direct entry into the holding pattern. Adjust time on the outbound leg to achieve one minute on the inbound leg.**

trolled Approach radar) is available, you may obtain its services by requesting a GCA approach. Or, they will monitor any other type approach if you request them to do so.

When you begin your approach, report leaving your last assigned altitude. Report the procedure turn, and report crossing the "gate," or low station inbound to the field. When you break out of the clouds, report that you are "contact." Don't report "contact" unless you definitely have the ground in sight, and can maintain flight with reference to the earth rather than instruments. Once you report "contact," your IFR flight plan is cancelled. You are expected to land as soon as possible consistent with safe flight procedures. Should you not become contact, you are required to report and begin the missed approach pull up procedure.

**General**

Detailed ATC procedures are given in the *Flight Information Manual* which is available from the Government Printing Office. There simply is no substitute for this manual—every instrument pilot should subscribe to it.

Keep in mind that ATC is a big organization, but that you are talking to *one* ATC controller who is just another human being with human failings, and human capacities. He cannot outguess you, outsmart you, or be any nicer to you than you are to him. If you don't know proper procedures, if you act like a smart aleck, file flight plans in the air when you could have done so on the ground in person or by phone, or otherwise make yourself a burden to the ATC controllers, they will still do their best for you, but—with teeth tightly clenched.

By law the controller is required to provide you with all the services he can give. He will do so even under the most trying circumstances. Try to realize his job is difficult at best, and becomes horrible when a pilot smarts off, or pulls a bonehead maneuver. I have never known an ATC controller who didn't have a pretty even disposition, a quick smile, and a pretty good outlook on life. I have never known a grouchy one, although I'm sure there must be some around. I have never known one who wasn't dedicated to his job in about the same manner as a brain surgeon. But they're still human. They can make mistakes. They can (and do) get tired, irritable, and short-tempered. You can help them by knowing the proper procedure, by *not* being a blabber-mouth on the radio, and by saying "thank you" occasionally. It won't break your jaw, and it just might get you preferential treatment.

Of course the big boys in Washington would do handsprings if they thought an ATC controller ever gave anyone preferential treatment. But the simple fact is, controllers are human—from the tower operators, to the FSS boys, to the ATC men in the Center. If you're nice to them they just can't help being nice to you. Try it and see!

Visit your local tower, FSS, and ARTCC. Spend at least two hours in each facility. Tell the men on duty you are a pilot, and want to learn as much as possible in the time you're there. Unless the traffic is heavy, they'll give you a grand tour with full explanations, and in terminology you understand. You'll learn things they don't write in books, regulations, and manuals. Ask questions, and listen to the answers. My personal opinion of this sort of thing is that until a man (or woman) has spent two hours in a tower, FSS, and ARTCC, he or she is not qualified to hold an FAA Instrument Rating. My students do it, or I don't recommend them for the rating!

Before visiting a facility such as the tower or FSS, call on the phone. Tell them your name, why you want to visit them, and how long you expect to stay. Ask if it's convenient for you to visit them at that time. If they say no (which they will when traffic is really heavy), ask for a time to visit. They'll be happy to tell you because they're much more anxious to have you visit them than you are to see them. They like to show pilots the controller's side of the picture, and for good reason. It helps both do a better job.

Where controllers are concerned the pilot has the upper hand—to a degree. Controllers are bound by stiff regulations which prevent smarting off to pilots, or giving us any guff at all. They're required to do their best even though they don't want to. Seldom do they get an opportunity to turn the tables, but I know one who did. A certain airline captain had been giving a certain ATC controller a bad time. The controller had the airliner on radar, and told him he was three miles right of the airway centerline. The captain, using his most sarcastic tone, replied, "I understand the federal airways are ten miles wide." (Which they were at that time.) There was a pause, then the controller said, "Yes sir, that's right, but the *professionals* fly on the centerline!"

"Do unto others as you would have them do unto you" has no better application than in the pilot-controller relationship. The more the pilot knows about the controller's job, the better able he is to see why certain procedures are set up the way they are, and the better able he is to see why a little graciousness goes a long way. The man on the ground can't fly the airplane for you, but he can come so close to doing so in an emergency that you'll think he's right there in the cockpit with you. It's a two-way street—try to get some experience on both sides of it!

**Lost-Communications Procedures**

If your aircraft is properly equipped for IFR, common sense dictates that you have at least two radios on which you can hear ATC controllers. Should one of them stop operating, you can use the other. Should both fail, you have a real problem, but not one that can't be solved. Detailed lost-communication procedures (radio failure procedures) are contained in the *Flight Information Manual,* and should be followed to the letter. They will not be re-stated here because of their length. It's a good idea to read them over, and become generally familiar with them although it is not necessary to memorize them.

Generally, the procedure is to maintain your cleared route of flight at the last assigned altitude, or the minimum enroute altitude, whichever is higher. Do what your ATC clearance says you are cleared to do. When you arrive at destination, you are expected to begin an approach at the expiration of the estimated time enroute shown in block II of the FAA Flight Plan (FAA Form 398). If you arrive on time, or late, immediately begin an approach. If you arrive early, it would be prudent to hold until your estimated time enroute is up, then begin the approach. Make calls in the blind on the radio on the chance that you are getting through, but can't receive anything.

If you have never suffered a loss of communications while on an IFR flight in the soup, then you're in for a thrill. I did once, and I can tell you it wasn't any fun! Keep your radios in good repair. There is no excuse for not having them in excellent condition when your life may depend on how well they work. When it happened to me the radios were in fine shape, but the radio installation on the aircraft I was flying was extremely poor resulting in frequent failures. One of them occured when I was in the soup on an IFR flight plan. The failure was due to rain leaking into various radio system components while in flight. Water is a good thing when you're hot and sweaty and need to take a shower, but it isn't worth a damn inside a radio set!

# 8. Planning an IFR Flight

Planning an instrument flight is no different from planning a VFR flight with the exception of the additional skill required to fly instruments. The more a flight is planned, and the more detailed information there is available to the pilot before the flight begins, the easier the flight will be, and the safer it is likely to be. To illustrate, we are going to plan a flight from the Fulton County Airport in Atlanta, Ga. to the Macon Municipal Airport in Macon, Ga. Although a very short flight, it will illustrate the procedures for *any* IFR flight.

**Route of Flight**

We are not going by the most direct route for the simple reason that we want the flight to last at least one hour. Therefore, we have planned the following route of flight:

Takeoff from Fulton County airport, fly direct to the Atlanta VORTAC, then Victor 97 to the Grant intersection, Victor 243 to the Pratt intersection, then Victor 56 to the Macon VORTAC.

Upon arrival at Macon we will make an ILS to the Lewis B. Wilson Airport. We will need to carry all of the approach procedure charts for the Lewis B. Wilson Airport as well as the charts for the Columbus Metropolitan Airport at Columbus, Georgia which will be our alternate.

**Materials Needed**

The charts and equipment we will need for this flight are as follows:

*Enroute Low Altitude Chart* for the route of flight (Chart L-20)

The Atlanta *Area Chart* (On Area Chart A-1)

*Instrument Approach Procedure Charts*

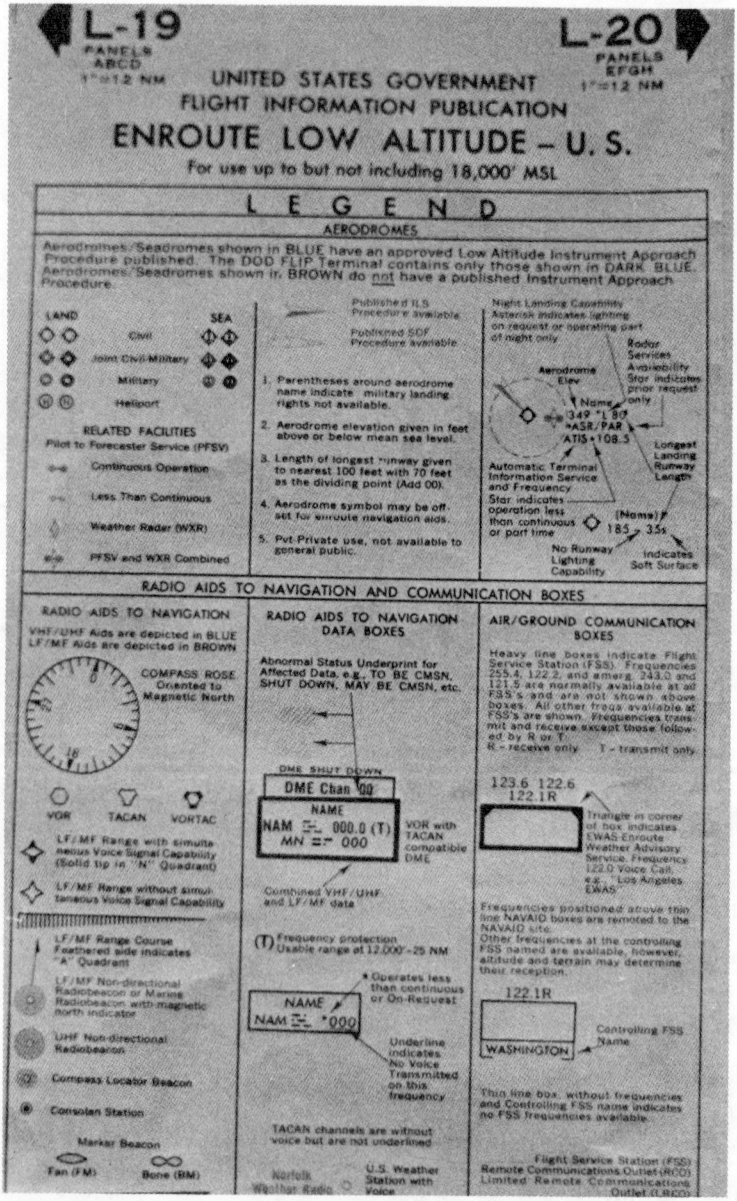


**Enroute low altitude chart legend should be familiar to every instrument pilot. Its use in planning stage of an IFR flight can be very helpful.**

The Atlanta Sectional Chart
A flight log form (Either one you devise or one published by any number of aviation publishing houses)
A copy of the FAA Flight Plan you file for the flight
A sharpened pencil and spare pencil
Current issue of Part 3 and 3A of the *Airman's Information Manual*
A rate-time-distance computer (E-6B or equivalent)

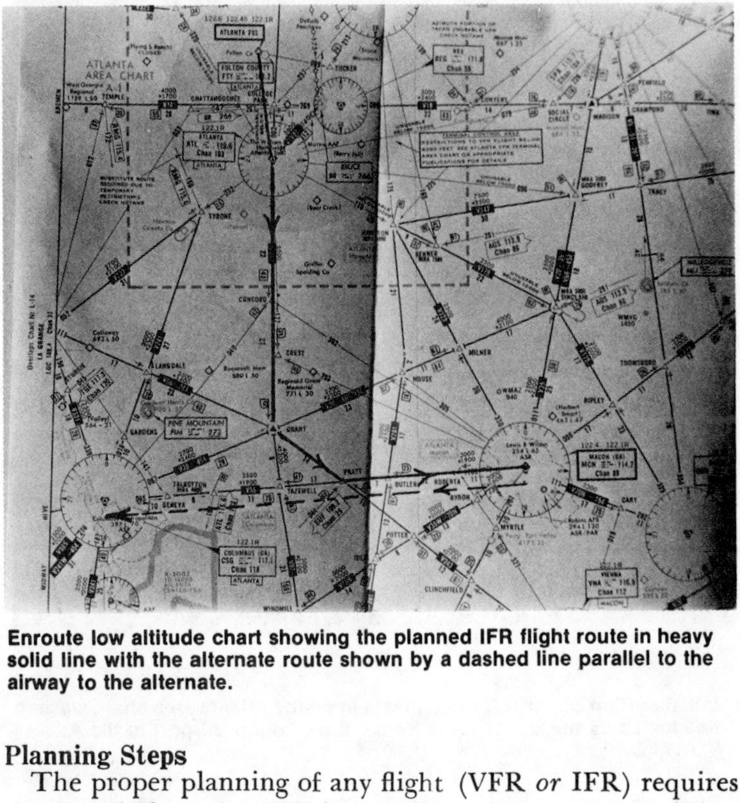


**Enroute low altitude chart showing the planned IFR flight route in heavy solid line with the alternate route shown by a dashed line parallel to the airway to the alternate.**

### Planning Steps

The proper planning of any flight (VFR *or* IFR) requires the completion of a definite number of steps, the omission of any one of which will seriously jeopardize the safety of the flight. The following steps indicate the necessary items to be accomplished:

1. Determine the route of flight (we've done this).
2. Assemble necessary charts, manuals, and equipment

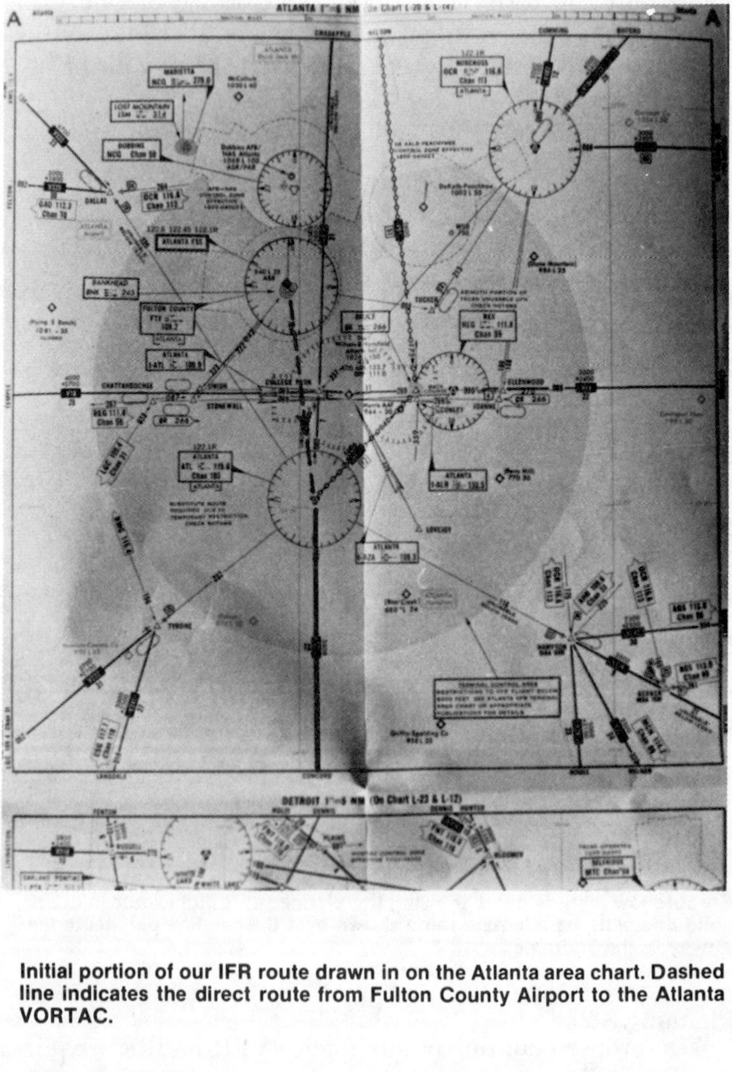


**Initial portion of our IFR route drawn in on the Atlanta area chart. Dashed line indicates the direct route from Fulton County Airport to the Atlanta VORTAC.**

(see the list).

3. Make the basic flight log with checkpoints, distances, headings, etc.
4. Check the weather – departure weather, enroute weather, landing weather, and weather for an alternate

airport of landing should one be required.

5. Check the Notices to Airmen (NOTAMs)
6. Obtain the winds aloft and complete the flight log.
7. File the FAA Form 398 (Flight Plan), and make a carbon to carry with you on the flight.

We've already determined our route of flight (Item 1), and made a list of charts and equipment (Item 2). Now, we need a flight log (Item 3). A number of flight log forms are commercially available, usually in a pad form. There is one on the reverse of the FAA Form 398 (Flight Plan) which is adequate for IFR flight. However, we prefer our own flight log, and it will be illustrated in this problem.

## The Flight Log

To begin our flight log, we need courses, checkpoints, and distances. These items we obtain from the various charts—Sectional and Enroute—IFR Charts. Our first leg is from Fulton County Airport to the Atlanta VOR so we check the Atlanta Sectional for the course and distance from the airport to the VOR station. The true course is 176° and with zero variation this makes the magnetic course also 176° which we enter in the "Mag Crse" column of our flight log. We measure the distance as 17 nautical miles and enter this on our log for the "leg" distance.

We measured our first leg on the Atlanta Sectional Chart, but now (from the Atlanta VORTAC) we are flying on an air way so we go to the IFR enroute chart for the remainder of our flight planning. The second leg is from the Atlanta VORTAC to the Grant intersection. Going to IFR enroute chart L-20 we find the distance from the Atlanta VORTAC to the Grant intersection to be 43 nautical miles. The magnetic course for Victor-97 is 179°. All distances on the IFR enroute charts are in nautical miles and all courses/headings are magnetic. Keep in mind that winds aloft are given in degrees true, so the winds must be converted to magnetic (east is least and west is best!) before the computer is used to determine groundspeed and heading.

At Grant intersection we will depart Victor-97 and fly

DATE 2-16-73 | AIRCRAFT CALL SIGN CESSNA 5811G

NOTES: V= VARY E= ESTIMATE CAS= 100K

TOTAL DIST. 104 TOTAL ETE 1+13

| CHECK POINTS | MAG CRSE | WIND | MAG HDG | LEG / REMAIN | LEG ETE / REMAIN | ETA / ATA | GS | TEMP / ALT | TAS |
|---|---|---|---|---|---|---|---|---|---|
| ATLANTA VORTAC | 176° | V | 176° VARY | 17 / 87 | :13 / 1+00 | | E 80 | V / 5000 | E 80 |
| GRANT ▲ | 179° | 210° 15K | 183° | 43 / 44 | :27 / :33 | | 95 | +10 / 5000 | 109 |
| PRATT Δ | 124° | ↓ | 132° | 14 / 30 | :08 / :25 | | 106 | +10 / 5000 | 109 |
| MACON VORTAC | 086° | ↓ | 093° | 30 / 0 | :15 / :10 | | 117 | +10 / 5000 | 109 |
| APPROACH & LANDING | V | V | V | — / — | :10 / 0 | | V | V / V | V |
| ALTERNATE: | | | | | | | | | |
| COLUMBUS VORTAC | 266° | 190° 20K | 256° | 70 / 0 | :40 / TOTAL 1+53 | | 105 | +8 / 6000 | 111 |

**Flight log designed by author. Any similar flight log will work so long as it shows essential information. The important thing is to plan well and thoroughly and have all the necessary information available before takeoff.**

down Victor-243 to the Pratt intersection. From the chart we determine this short leg to be 14 nautical miles and the course is 124°. Our final leg is from Pratt intersection to the Macon VORTAC and the distance is 30 nautical miles on a course of 086°. We now have planned our flight from departure to the fix (Macon VORTAC) from which we will make the approach. Our next entry on the log is for the approach and landing. We will allow ten minutes for the approach and landing and enter it on the log sheet simply as "approach and landing." We will consider it for fuel planning purposes as ten minutes at cruising flight. We now add up the column for distance beginning at the bottom and addding upward. Doing it this way gives us the distance remaining, and when we apply the wind and determine the groundspeed we will compute the times for the flight in the same manner, i.e. we will show leg times and times remaining.

**Checking the Weather**

The most common method of checking the weather is to call or visit the local Flight Service Station. In many cities you can also call the U.S. Weather Service. The difference is that the Flight Service Station has a "Weather Briefing Specialist" who can only read the existing and forecast weather to you while the U.S. Weather Service has a forecaster who can give you his expert opinion about the weather. If you have a choice, call the U.S. Weather Service. Be sure to tell him you want an aviation forecast.

When checking the weather the pertinent items to check are the existing and forecast ceilings and visibilities at your destination and alternate, the freezing level, winds aloft, and any unusual weather along your route. You should pay particular attention to any forecasts of freezing rain (don't go unless your airplane is well equipped with anti-icing equipment) and icing. If thunderstorms are forecast you should not go unless your airplane is equipped with airborne radar.

**Check the NOTAM (Notices to Airmen)**

A NOTAM file is maintained in the FSS. We should check it to see that all radio aids to navigation for our route are working properly. Should any be shut down, a NOTAM will be issued for it. NOTAMs are issued on shut-down or malfunctioning radio aids to navigation, and are issued for any dangerous situation that might affect safety of flight.

We should also check the NOTAMs published in the *Air-*

*man's Information Manual,* and the Special Notices in the *Airman's Information Manual.*

Many pilots try to fly in thunderstorms and rely on Air Traffic Control Center radar. This is not wise. ATC (Air Traffic Control) radar is primarily for that purpose — air traffic control. In order to see the traffic (airplanes) the radar used by ATC employs MTI (Moving Target Indicator) and CP (Circular Polarization). MTI means that any target that is not moving will not be shown on the radar scope and CP is used to remove precipitation targets, i.e. rain, snow, clouds, etc. The normal configuration of ATC radar will be with MTI and CP in use. With these two functions operating the ATC radar will show very little weather. It is very unwise for a pilot to rely on ATC radar to vector him around storms.

Airborne radar, on the other hand, is designed to show weather, and properly tuned, will give accurate indications of the location of storms. Anyone who plans extensive IFR flying should have an airplane equipped with airborne radar.

For the flight we are planning the weather is a warm front lying north and south roughly between Atlanta and Macon. Ceilings in advance of the front are 300 to 700 feet with rain and low visibilities. To the west conditions are improving and Columbus, Georgia has 3,000 feet with a visibility of 7 miles and is forecast to improve. We will use Columbus as an alternate and enter on our flight log the distance and course there (70 nautical miles and 266°).

**Completing the Flight Log**

Our next step (Item 6) in planning the flight is to obtain the winds aloft, and complete the flight log. Flying eastbound on the airways we should file for an odd-thousand-foot altitude (3,000, 5,000, etc.). Flying westbound, we should pick and even-thousand-foot altitude (4,000, 6,000, etc.). Eastbound is a magnetic course from 360° to 179°. Westbound is from 180° to 359°. Since all of our flight is eastbound (magnetic course from 360° to 179°) we will use an altitude of 5,000 feet. We next check the IFR enroute charts to be sure this altitude is above the MEAs for our flight.

The weather forecaster gives us a winds aloft for our flight of 210 degrees at 15 knots, and a flight level temperature of +10°C. We enter these items on our flight log, and now can determine the groundspeed and times. Our first leg from the Fulton County Airport to the Atlanta VORTAC is only 17 nautical miles, so we will be climbing during this leg. We estimate 80 knots as our groundspeed and figure the wind will be variable. For climbs to very high altitudes use the average true airspeed in the climb and the wind at two-thirds of the final altitude. For our flight the estimated climb speed will be accurate enough.

Now we use the computer (see *Computer Guide* by Frank Kingston Smith) to determine the times for each leg of the flight, and we enter these on the flight log. Include the time to the alternate (Columbus, Georgia) and add this to the flight time to determine the total estimated time enroute for fuel planning purposes and to ascertain that you have enough fuel in reserve to meet the IFR fuel reserve requirements (time to destination, plus time to alternate, plus 45 minutes).

The final planning step is to complete the FAA Flight Plan form, and file it with the local Flight Service Station. Ideally this should be done in person, but if you are not at an airport with a Flight Service Station, use the telephone and call it in. Before you make the phone call, have the flight plan form filled out — this will conserve both your time and the time of the Flight Service Station specialist, and will serve to get your flight plan into the ATC system quicker.

When you file an IFR flight plan with an FSS they transmit it to the air route traffic control center (ARTCC) which you will be flying in. The ARTCC makes up a flight strip on your flight and checks traffic to see when they can allow you to take off and enter the system. They then issue a "clearance" and normally this comes to you from the tower. If you are at an airport without a tower you receive it from the local FSS. Should you be at an airport where there is neither a tower nor an FSS you would have to take off under VFR conditions and receive your IFR clearance in the air from the nearest tower or FSS.

When you receive the ATC clearance you are not required to read it back unless requested to do so. But, be sure to copy it carefully and note any differences from the route or altitudes that you filed.

**The Flight Kit**

We have already listed the materials needed for our IFR flight. In addition we should have some kind of bag or briefcase in which to carry our materials. Except for the maps and charts, the flight kit and other materials need not cost more than $50 to $75 depending on the type you buy. If this seems excessive, remember that a kit bag (flight kit), plotter, computer, etc. are life-time purchases if you care for them, and don't leave them sitting in some airport office!

Although we did not include it in our list of materials needed for an IFR flight, we highly recommend a knee board. Several are commercially available, and the cost is nominal. Your flight log, flight plan, and the chart you are using can be clipped on the knee board. This prevents dropping them on the cockpit floor, or having them blow out a window should you open one in flight.

A flight kit and knee board aren't essential—you *can* go out to the airplane with maps and charts jammed in your pockets, a plotter stuck in a rear pocket, and a pencil between your teeth. If you make it to the airplane without losing anything stuff it all on top of the instrument panel, or on the floor beside your seat. Once in flight, you'll have a devil of a time finding what you need, but at least you didn't lay out five to ten bucks for a flight kit and about the same for a knee board! If you *do* buy a flight kit and knee board, place the flight log and flight plan on the knee board along with the first chart you plan to use. Put the remaining charts in the flight kit in the order you will use them. Then, when you need them in flight, they'll be easy to find.

If you're a student contemplating the instrument rating, or if you're a private or commercial pilot and the instrument rating is your next goal, do yourself and your flight instructor a favor. Before you call the flight instructor for that first lesson, buy a good E-6B or Jeppesen computer. One side is a slide rule for rate-time-distance problem solution, and the other side is the wind face for solution of wind vector diagrams. They are made of either plastic or metal, and come in two sizes—standard and shirt-pocket size. Except for the size, the standard and pocket size computers are identical.

Buy a good plotter. They are made of plastic, and are used for measuring courses and distances on aeronautical charts. Every serious instrument pilot should have a plotter and computer in his flight kit. He should also purchase (and read) a book on instrument flying—which you have quite obviously already done if you're reading these lines! Buy all the necessary maps and charts, and assemble a flight kit. When you've done all this, *then* call on an instrument flight instructor for your first instrument flight lesson.

# 9. Flying an IFR Flight

In Chapter 8 we planned an IFR flight from the Fulton County Airport in Atlanta, Georgia direct to the Atlanta VORTAC, victor-97 to Grant intersection, victor-243 to Pratt intersection, victor-56 to Macon. We listed Columbus, Georgia as our alternate. It is wise to *always* plan an alternate, even when your destination weather forecast doesn't require it. Keep in mind — poor weather is more difficult to accurately forecast than good weather.

The last step we performed in Chapter 8 was to complete and file the FAA Flight Plan (FAA Form 398). This can be done in person at an FSS, by telephone to an FSS, or by radio if there is absolutely no opportunity for you to use a telephone prior to takeoff to file the Flight Plan. Pilots who file by radio after takeoff are within their legal rights and no one can stop them. But filing in the air clutters up the radio, and is inconvenient to everyone except the pilot doing the filing. It is also very un-professional. Of course, there are times when it is impossible to file a flight plan prior to takeoff, and it is for those times that filing in the air is acceptable. But, before you take off and file in the air, be sure you can't file in person or by telephone.

Once you file the Flight Plan, the Air Traffic Control Center assumes you will go directly to your airplane, start the engine, and call the tower for taxi instructions. At some large terminal areas you may experience up to a thirty-minute delay in obtaining a clearance. If this situation exists, the FSS with whom you file will notify you that the delay may be expected, and you can have a cup of coffee before you crank up and taxi out. However, if they do not mention a delay, then you should go directly to your aircraft and prepare to start the engine.

For this reason it is always a good idea to accomplish the aircraft preflight before you file the Flight Plan. Then, once the Flight Plan is filed, all you have to do is start the engine and go. The aircraft preflight warrants a word or two at this point. The FAA regulations specify minimum equipment for actual IFR flight, and you should ascertain that your aircraft is so equipped, and that all of the required equipment is working properly. Check any anti-icing or de-icing equipment for proper operation. Check the radios while taxiing out. Don't forget the VOR receiver must be checked for accuracy every ten days or ten hours of flight, whichever occurs sooner. Use the VOR test facility or ground check points, or both.

**VOR Receiver Test Facility and Ground Check Points**

VOR receiver test facilities are located at various airports around the country, and are used for checking accuracy of VOR receivers. The radiated test signal of the VOR receiver test facility transmits one radial at all points around the station. Thus, any aircraft tuning the radiated test signal receives one radial. The procedures for using this radial to check your VOR receiver are outlined in the *Airman's Information Manual.* For most receivers you set 180° on the course selector, and the "To-From" ambiguity meter will read "To" with the CDI (Course Deviation Indicator) needle centered. If the CDI is not centered, your set needs adjusting. If the "To-From" meter does not indicate "To," your set needs to visit a radio repair shop.

Ground check points are established at airports where a nearby VOR station may be clearly and accurately received by an aircraft on the ground. These points are published in the *Airman's Information Manual,* and are shown on the airport by a sign near the spot where the VOR ground check may be made. The sign will contain the frequency of the VOR station, and its identification. The radial and/or course will be shown, and the proper ambiguity meter reading will be given. Park by the sign, and tune the VOR. Your CDI should be within plus or minus 4° of the reading shown on the sign. If it is not, your set may need adjustment.

**Equipment For IFR Flight**

At the present time, the FAA requires the following items for IFR flight in a private aircraft not being flown for hire:

Airspeed Indicator
Altimeter (sensitive, with barometric adjustment)
Magnetic Direction Indicator
Tachometer (one for each engine, if twin)
Oil Pressure Indicator
Coolant Temperature Gauge if liquid-cooled engine
Oil Temperature Gauge if air-cooled engine
Manifold Pressure Gauge if engine is blown
Fuel Quantity Gauges
Approved Safety Belts
Approved Navigation Lights (if night IFR)
Anti-Collision Beacon (on large aircraft)
Sufficient electrical power for all equipment
Two-way Radio for Communications, and Navigational Equipment appropriate to the ground facilities to be used.
Gyroscopic Rate of Turn Indicator (turn and slip)
Ball Bank Indicator (incorporated in turn and slip)
Clock with sweep-second hand
Artificial Horizon
Directional Gyro (or equivalent)

**Cessna 150 instrument panel is simple and neat. Although FARs do not specify minimum number of radios that must be installed for IFR flight, experienced pilots want at least two communications radios and two navigation radios. (Cessna Photo)**

Note that, on the required radio, they only specify 1) Two-Way Voice Radio, and 2) Navigational Equipment appropriate to the ground facilities to be used. Since our federal airways are based primarily on VOR, it can be assumed that a basic VOR receiver is adequate to serve this requirement. However, it is highly desirable that your VOR receiver be able to receive the ILS systems, and that you have a marker beacon receiver. The old Narco Omnigator Mark II is a very inexpensive radio that will serve the basic requirement of two-way voice radio, and sufficient navigational equipment. However—and this is a strong *however* —be prepared for a very rough time of it! The set must be hand tuned, and trying to use a hand-tuned set (instead of a crystal-tuned one) and fly an airplane IFR is about like trying to balance an egg on your nose while riding a bicycle. It *can* be done, but most pilots won't advise it. Besides, there are many good "one and one-half systems" available at not too great a cost. A one and one-half system is a complete voice transceiver (transmits and receives on same frequency) which is crystal controlled, and a VOR/ILS receiver that is also crystal controlled. Usually this entire set mounts on one panel with the CDI and ambiguity meter in the center. A Narco Omnigator Mark II is an excellent second VOR set and back-up voice radio for a one and one-half system.

There is no specific requirement to have an ADF set in your airplane for IFR flight. An ADF set is nice to have, and helpful when making an ILS approach. It will be the major navigation set if you plan to fly overseas, or into Mexico or Canada where low frequency stations are more prevalent than they are in this country. But, for normal flying of the airways in this country, the ADF set is not absolutely required. We like to have one, but we know pilots who have flown IFR for several years without one. It boils down to a matter of choice, and of how much money you have to spend. Certainly, the VOR equipment you need comes first since you can't fly without it on our federal airways.

**Getting Airborne**

The most dangerous portion of any IFR flight is from

your home to the airport if you go via automobile. It's safer if you walk, but it usually takes longer. In any case, the first step is to get to the airport. Once you arrive there, the next step should be to preflight the aircraft, and ascertain that it is in good working order, and that its equipment is also in good working order. Once this is assured, arrange the materials needed for your IFR flight in a kit bag so they will be easy to reach in flight, and go check the weather. The approach charts show minimum weather you must have at destination and alternate. You check the weather to be certain it is at least as good as the approach charts require.

It is almost impossible to say *too* much about checking the weather. Weather figures as a primary or contributing factor in the majority of general aviation accidents—pilots fail to check it, or fly anyway when it's bad, or fly their aircraft into weather neither they nor the aircraft are equipped to be in. Thunderstorms are simply to be avoided. It is not economically feasible to build a general aviation aircraft with sufficient structural strength to make thunderstorm penetration a routine matter. For most airplanes and pilots, the sensible and safe route to follow is *away* from thunderstorms.

If your aircraft is not fully anti-iced, don't file into known or *suspected* icing conditions. Don't try to fly in known icing conditions (or even suspected icing conditions) unless your aircraft has wing and empennage anti-icing or de-icing equipment, a heated pitot tube, and propellers with some form of anti-icing or de-icing. Don't expect ATC to give you an ice-free altitude if your airplane doesn't have anti-icing equipment. Air Traffic Control is strictly a matter of "first come, first served" without regard to who you are, or what you're flying. They don't have the time, nor is it feasible, to allow you to make random changes in altitude to get out of icing conditions. So, if the forecaster says there'll be a chance of icing, and your airplane isn't equipped to handle it, stay on the ground!

**A Typical Flight**

We flew this flight solo in Fulton Air Service's instrument trainer, a Cessna 150, N5811G. Let me give you a little advice on flying solo IFR (or with passengers and no copilot). It is legal, but not always good sense. An autopilot makes a pretty good copilot IF you have practiced using it and can take maximum advantage of it. I have flown a lot of single-pilot IFR, but it takes good planning and not a small amount of practice. I made a lot of mistakes and came close to having an accident on more than

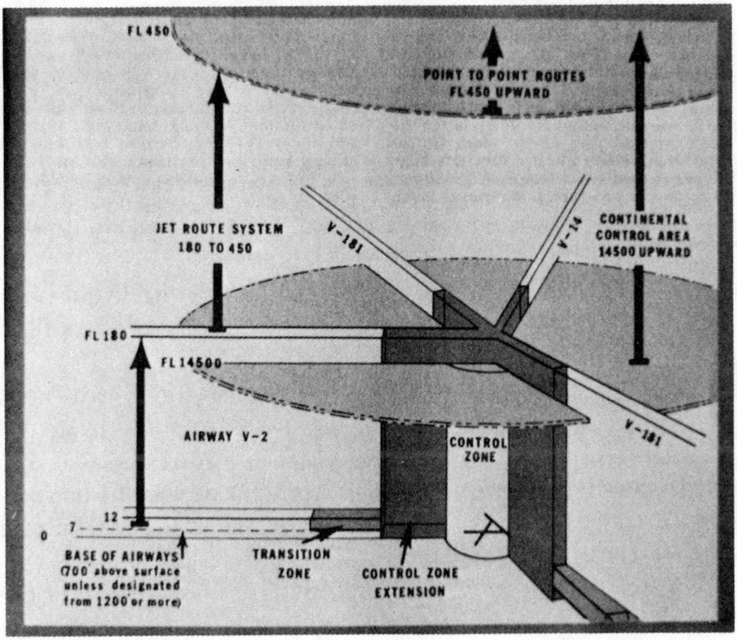


**Simplified diagram of the airway structure may help you visualize its construction.**

one occasion. If I had it to do over again I'd take along an experienced IFR pilot as copilot until I had enough time in the actual soup to feel comfortable.

Which brings up another point. Actual IFR is nothing like flying under the hood! All that gray stuff around the cockpit won't go away! It's frightening the first time you see it, and if you're also solo with no IFR rated pilot to help, things can go to hell in a handbasket quicker than you can

**The Atlanta Flight Service Station. Pilot briefing facilities are in the foreground. Controllers in the background transmit flight plans, give assistance to pilots in flight. Flight Service Stations are essential to the general aviation pilot. No where else can he obtain professional help and guidance in planning a flight, either VFR or IFR.**

imagine. And don't make the mistake of thinking a non-IFR rated copilot will do — they sometimes are less than worthless — more of a hindrance than a help.

And don't forget the impact your passengers can have. A frightened passenger isn't funny any time and in actual IFR the situation can get dangerous in a hurry. No need to point out how troublesome an airsick one would be. Better your initial experience in actual IFR should be just yourself and an IFR rated copilot and no passengers.

Rather than give you a blow-by-blow description of the flight in Cessna N5811G I'd like to take the remaining pages of this book to discuss a few of the very practical aspects of actual IFR flying. First of all, until you've got a lot of actual IFR hours in your log book, it's frightening. Show me a new instrument pilot who isn't firghtened and he's either not telling the truth or just too dumb to know better. Good, safe pilots are not fearless men. The airline pilot who said airline flying was hours and hours of boredom punctuated by brief moments of stark terror was telling it pretty much like it is.

Fear is normal and everybody experiences it — Wilbur and Orville did, Lindbergh did, John Glenn did — every

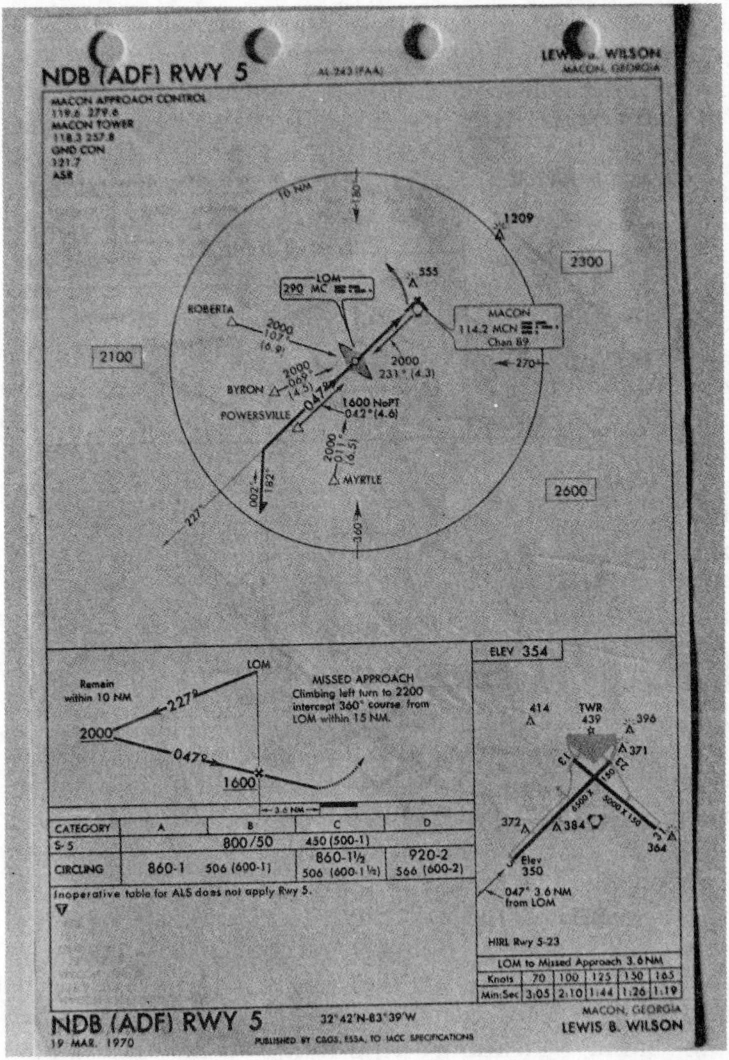


| CATEGORY | A | B | C | D |
|---|---|---|---|---|
| S-5 | | 800/50 | 450 (500-1) | |
| CIRCLING | 860-1 | 506 (600-1) | 860-1½ 506 (600-1½) | 920-2 566 (600-2) |

| LOM to Missed Approach 3.6 NM | | | | | |
|---|---|---|---|---|---|
| Knots | 70 | 100 | 125 | 150 | 165 |
| Min:Sec | 3:05 | 2:10 | 1:44 | 1:26 | 1:19 |

**NDB (Non-Directional Beacon) approach to the Macon, Georgia airport using an ADF (Automatic Direction Finding) procedure. This is a non-precision approach procedure in that no electronic glide slope information is available. All NDB and VOR approaches are non-precision approaches and the pilot descends to the prescribed MDA (Minimum Descent Altitude) and flies his time from the final approach fix to the missed approach point. If he sees the runway he may land. Otherwise he must execute a missed approach.**

professional pilot has, at one time or another, had the living daylights frightened out of him — and all of them experienced fear in actual IFR when they first started. Fear is okay, but panic is verboten. You can't panic — you've got to stay in there and fly the machine. It takes conscious effort sometimes, and it becomes a real chore and you wish to hell it was all over and you were back on the ground, but you have got to fly the machine — all the way to the parking spot. Nobody can do it for you and all the air traffic controllers in the whole world can't keep your air machine upright — *you've got to!*

It should go without saying that the more good practice you've had, the better you'll be able to handle the situation. That's one reason I advocate the IFR rated copilot until you are pretty sure you are comfortable when it's all gray outside the cockpit. Remember, nobody was ever *too* good at doing something, but a lot of pilots haven't been good enough.

Know the system. I mean the IFR and ATC system. Visit the ATC facilities from FSS to tower to approach/departure control, to the center. Don't just stand around and look — ask questions and get them to tell you how it is from their side. The more you know the system the better you'll function in it. When you're out flying the system, listen to what they're telling the other airplanes and try to visualize traffic conflicts — controllers are human and they can err.

Back in the days before there was much radar coverage I was flying down an airway in the soup and there was a faster airplane behind me at my altitude. His ETAs to the next checkpoint kept getting closer and closer to mine and finally he gave them one that was one minute *ahead* of mine! That got my attention, but apparently the controller didn't mind. I mentioned it to him, and there was a moment of silence. Then, he called the airplane behind me and had him climb immediately to a higher altitude and his voice had strong emphasis on the word "immediately". I have no proof that the controller had missed the point that, on the leg we were flying the airplane behind me would pass me. But, I do know that I would never have

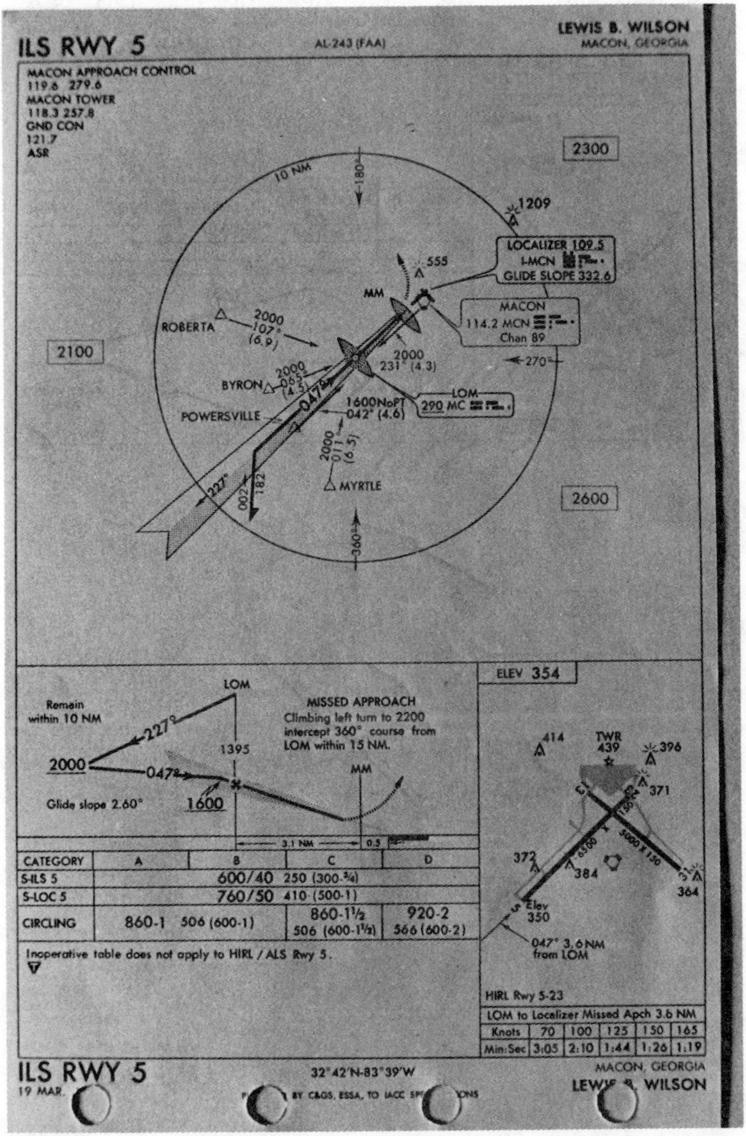


ILS (Instrument Landing System) approach to the Macon, Georgia airport. This is a precision approach as electronic glide slope information is available if the airplane has a glide slope receiver. Pilot descends down glide slope to the DH (Decision Height) and if contact he lands, otherwise he must do a missed approach.

been able to mention it to him had I not been paying attention to the other pilot's ETAs.

A good friend of mine flew into the side of a California mountain because a controller misidentified him on radar, and my good friend wasn't keeping track of where he was. He blindly followed the controller's instructions. The tragedy of it was, he must have begun to realize something was amiss because his last transmission was "I don't think you've got me — I'm getting the hell out of here." Seconds later he hit the mountain.

Another friend hit a four thousand foot hill in New York. Same reason. Blindly following radar. A third friend just didn't know for sure where he was and let down early. He hit a two thousand foot hill just a few feet below the top.

The name of the Big Game is to know, for sure, where you are — radar is great, but don't blindly follow it. They make mistakes and when they do, you're the guy who has to pay. And the price is VERY high!

You know something else IFR pilots do that's really dumb? They go below minimums — every pilot with an IFR rating sooner or later is tempted to try it. I've been there — none of us is guiltless. I've done it, but I won't do it again — the real problem is that you can get away with it so many times before you get caught, and then it's too late — the Big Game is almost always over. Death has a way of being extremely final.

The minimums are based on certain very specific criteria, and they take into account many variables — some of which many of us are not even aware of. Many pilots think the minimums are super conservative — let me assure you *they are not!* They are just what they say they are — minimums. And here is one place where the FAA doesn't pad the figures with any fat — when they say altitude "X" is the minimum, baby you can be sure they mean just that!

When you start fudging on the MDA (Minimum Descent Altitude) or the DH (Decision Height) you are messing around with pure fire. The sick thing about it is that many pilots who make it below minimums once or twice get confident in doing it — it's a false confidence that will sooner or later spell disaster. Better you should hunt tigers with a B-B gun!

Don't mess around with thunderstorms. If you have airborne radar you can perhaps pick your way through them, but you'd better know your radar set and you'd better keep a way out. NEVER depend on the ground radar — they use MIT and CP and will see only the most intense storms, if they see anything. And don't listen to anyone who tells you your airplane can fly "through anything" — airplanes come apart in thunderstorms. It is possible to fly into a thunderstorm and come out the other side in one piece — many of us have done it. But nobody can tell you how to tell a thunderstorm whose intensity is flyable without structural breakup from one whose intensity will destroy any airplane.

In twenty-two years of flying I have been in two thunderstorms — neither of them intentional. Once, in a flight across the Pacific I flew 200 miles off course to go around a line of them because I wouldn't fly into one intentionally. Give thunderstorms a wide berth — they deserve it!

Stay out of icing. Heated wings, empennage surfaces, and good prop anti-icing boots will take care of some pretty rough icing, but none of the other stuff will. Inflatable boots will do a fair job in light icing, but you'd better be skillful in operating them. And remember the cardinal rule of weather forecasting — bad weather is more difficult to accurately forecast than good weather. No forecaster on earth can *guarantee* you light icing — if icing conditions are forecast, assume it will be heavy and will accumulate rapidly.

I can very vividly remember two flights on which light rime was forecast and the accumulation was three inches of ice in less than sixty seconds. A good friend of mine flying a medium size twin with inflatable boots got so much ice during the procedure turn of a VOR approach that he couldn't maintain flight and mushed into the ground. Fortunately he hit in a snow covered field and suffered only minor injuries. The ice accumulation was so fast that it covered the boots in the inflated position. I have had this happen myself, and luckily we were light enough to make it in — with climb power on both engines. The point to remember about icing is that, unless you have the very best anti-icing (not de-icing) equipment, plan to avoid it.

IFR flying is not all bad — otherwise those of us who regularly do it wouldn't! Being instrument rated is being a whole pilot. There are many instances when it's VFR that you can't see the earth's horizon and need the ability to fly instruments — like on a dark, overcast night, for example. Or a day when the haze is so bad it obscures the horizon. Or a night takeoff out over an unlighted swamp area or other uninhabited area with no lights.

Like flying aerobatics, the ability to fly instruments adds that final little touch to a pilot's finesse. It adds utility to the airplane and extends his horizons. It has, in many ways, its own rewards — like breaking out on top of a solid overcast into brilliant sunlight — like the time over the North Atlantic when I broke out of the clouds into a cathedral made of clouds above and below me, with columns of clouds like the columns in Westminister in London.

And lastly, the ability to fly instruments gives a pilot confidence. Over-confidence is fatal in flying — a quiet amount of sureness and confidence is admired and respected, and is very valuable to the pilot. It comes from a lot of practice, and the humble ability to realize none of us is the best, but we are no good if we do not strive to *be* the best! Good flying!